Joseph Franz

Figure 1—Joseph Franz (same as cover photograph)

Joseph Franz

A Renaissance Man in the Twentieth Century

Jo F. Humphrey

iUniverse, Inc.
New York Lincoln Shanghai

Joseph Franz
A Renaissance Man in the Twentieth Century

Copyright © 2006 by Jo F. Humphrey

All rights reserved. No part of this book may be used or reproduced by any means, graphic, electronic, or mechanical, including photocopying, recording, taping or by any information storage retrieval system without the written permission of the publisher except in the case of brief quotations embodied in critical articles and reviews.

iUniverse books may be ordered through booksellers or by contacting:

iUniverse
2021 Pine Lake Road, Suite 100
Lincoln, NE 68512
www.iuniverse.com
1-800-Authors (1-800-288-4677)

ISBN-13: 978-0-595-36046-8 (pbk)
ISBN-13: 978-0-595-80496-2 (ebk)
ISBN-10: 0-595-36046-7 (pbk)
ISBN-10: 0-595-80496-9 (ebk)

Printed in the United States of America

Contents

Introduction		vii
Chapter One	The Beginning	1
Chapter Two	Settling in New York	12
Chapter Three	Becoming Self-reliant	26
Chapter Four	The New American	37
Chapter Five	The Journey Home	45
Chapter Six	The Thousand-Dollar Table	56
Chapter Seven	Life-changing Experiences	61
Chapter Eight	The Saga of the Stockbridge Lighting Company	..71
Chapter Nine	A New Family and Social Life	77
Chapter Ten	Safety Insulated Wire and Cable Company	90
Chapter Eleven	Settling in the Berkshires	99
Chapter Twelve	Expansion	113
Chapter Thirteen	Power and the Roaring Twenties	125
Chapter Fourteen	The Healing	135
Chapter Fifteen	Power, Politics, and More	145
Chapter Sixteen	Family and Friends	154
Chapter Seventeen	Community Commitment	168
Chapter Eighteen	The Shed	183
Chapter Nineteen	The War Decade	198
Chapter Twenty	Jacob's Pillow	210
Chapter Twenty-One	The Twilight Years	220

Introduction

In December 1955, Joseph Franz (1882–1959) wrote the following:

"For some time, it has been on my mind to put papers and articles concerning my life in some sort of chronological order. It is hoped that this documentation will help my children and grandchildren understand my part in their heritage. In many ways, I, Joseph Franz, am a complex personality. Time and distance dim remembrances and diminish the ability to identify with me.

In my European education, "Das Werk muss den Meister loben" (a man's work must praise the master) was the rule. "Blowing your own horn" was considered very unethical. Nevertheless, as Dr. Gamme once said in a speech to the American Institute of Electrical Engineers in Pittsfield, Massachusetts, "It's a mistake. No one else will praise you if you don't do it for yourself." In America, this certainly seems to be true.

Therefore, before time runs out, I'm putting down some facts, thoughts, and recollections about life as I have lived it. In youth, we often take our heritage for granted with little or no appreciation of it. It is not until maturity that we have the capacity to realize the great gift that is ours. I was fortunate to have come from a family that encouraged my visions and promoted my ability to bring to successful conclusion whatever I set out to accomplish.

In November 1955, our son, Peter, suggested that I write a story of my accomplishments. And so, on December 4, 1955, I have started to relate "life" as I have known it."

Joseph Franz was a human being born with potential strengths and weakness that would develop over time. In the world he came from, emotions were rarely exposed. At home in America, he was able to express his love for his family. Eventually, he was also able to discuss his difficulties and disappointments with his second wife, Emilia. She was just as independent as he was and could understand his concerns as well as his needs. It is time for his story to be told!

It has taken almost fifty years for our father's wishes to become a reality. Perhaps it has taken that long for us, his children, to mature enough to undertake the monumental task of compiling this book.

This work is based on our father's writings, journals, and notes. It includes the personal memories of his second wife, Emilia Radell Franz; daughter, Natalie Franz Hewlett; son, Russell Franz; and sister, Barbara Franz Gumpinger. It also includes recollections from and other family members, and friends. To all of them, we are deeply grateful. Through their help, we have tried to flesh out the man as a scientist who helped refine and spread electricity as a primary source of light and power to industry and homes of the Guilded Age. His daring innovations also made the newly invented electric street railways safer. As a philanthropist he freely gave his time and expertise to preserve the natural beauty of the Berkshires and increase the cultural opportunities that make life more enjoyable. As a teacher he unselfishly shared his knowledge with people of all ages and took an active part as a concerned citizen in town affairs. And to his family he was a kind and loving husband and father.

We also deeply appreciate the assistance of Barbara Allen at the Stockbridge Library Association Historic Collection; Norton Owen, historian of Jacob's Pillow; and of William E. Wood, chapter director and publication manager of the Connecticut Valley Chapter, National Railway Historical Society. For her many hours rendering the photographs and text for publication, we give a very special thanks to Deryl H. Clune.

Joanna Franz Humphrey, Shirley Franz Miller, and Peter Franz
2005

Chapter One

The Beginning

Joseph Franz architect, electrical engineer, inventor, artist, local politician, and philanthropist arrived in America aboard the *Königen Luise* on October 16, 1897. Having turned fifteen the preceding July and already graduated from his engineering studies at the Vienna School of Polytechnical Industry in August, he felt ready to face the future in the new world.

Figure 2—Joseph Franz, age fifteen

Because there weren't any steerage passengers, the ship berthed in Hoboken.[1] To the impatient young man, it was an endless time before he could leave the ship. First class passengers had already disembarked. From the lower deck, he watched as they found their luggage and proceeded through customs. His turn finally came. He didn't have to look for any luggage. He only had one small satchel that he kept in his cabin. He carried it off the ship himself. His father had shipped his trunk directly from Vienna to his Uncle Fritz in New York City by freighter, so he didn't have to worry about hauling that from the ship.

Even though he had very little command of the English language, while aboard the ship, he practiced his answers for customs.

"Mein name ist Joseph Franz."

He had been born Bernhardt Josef. However, he wanted immediately to appear as American as possible. He was healthy, and, thanks to his father's careful preparations, his papers were all in order. The noise and confusion of the first few hours in America soon faded into the morning sun as he set out to cross the North River.[2] He followed the instructions his uncle had sent to take the ferry from Hoboken to Liberty Street on the West side of Manhattan.

The ferry ride offered a view of skyscrapers competing with the dwindling number of tall ships that were still docked along the waterfront of lower Manhattan. On shore in New York City, a cacophony of sounds greeted Joseph. Some were familiar, like the clatter of streetcars and horse drawn vehicles. Others were unfamiliar, such as the rumble of the overhead trains. Various sounds—church bells, factory whistles, and the screeching of steel brakes as trams and trains came to stops or stations—were also included. Trying to communicate over all of the other noises, people everywhere were shouting at each other in a language still largely unknown to Joseph.

Joseph had to get to the 3rd Avenue elevated train, which was several blocks across town. On the way, many new sights confronted him. The great, elevated train system fascinated him. The elevated train lines stopped at all of the ferry terminals, creating a tangle of tracks and pillars nearly obscuring the sky. In Vienna, transportation was only at ground level, similar to the street rail system and horse drawn conveyances he observed.

---

[1] Only steerage passengers had to go through Ellis Island.
[2] The North River is now known as the Hudson River.

The ride uptown above 3rd Avenue brought many amazing sights, specifically the skyline of lower Manhattan. Church steeples and the towering Telegraph and Insurance Buildings had risen to new heights. Some of the commercial and residential architecture was similar to that in Europe, but the building materials were different. Along the way, signs of construction for more buildings were evident.

Many German immigrants had moved to the East Side of Manhattan. At the 68th Street stop, Joseph left the train. When he descended to the street, he was amused to hear people speaking his own language. Street vendors selling roasted chestnuts reminded him of home. It wasn't hard to find the East 66th Street apartment of his cousin, Amalia Hettrich Fetzer. They had met when he was a child, so they greeted each other warmly. She knew the date and time of the boat's arrival. She guessed it would probably be around noon when Joseph arrived at her apartment. She presumed he would be hungry, so she had prepared a big meal.

During the afternoon, Joseph learned about all of the relatives who had preceded him to America. The older ones had already been educated or trained in a profession when they arrived, so it didn't take them long to become assimilated into a similar strata of society in their new surroundings. Both Amalia Hettrick and Fritz Franz had apartments in the predominantly German community of Yorkville on the East Side of Manhattan. Other relatives had found homes in New Jersey and the Bronx, where real estate was less expensive. All of the relationships were very confusing to the teenage Joseph, particularly because he was too young to have known most of these people in Europe. With so many faces at once, he began thinking he was related to everyone in town.

∗   ∗   ∗

Joseph was born into a German family that was part of a growing middle class of merchants, civil servants, and professionals by the mid-nineteenth century. Germany had finally become united. Prussian militarism was the driving force of the German industrial revolution. Individual ingenuity was encouraged, and the Franz family profited due to their own creativity and intelligence.

Joseph's grandfather, Karl Johan Franz, had invented the hollow concrete block, which enabled buildings to rise higher and allowed walls to support more weight with less bulk. Because of his success, he and his brother opened a thriving building supply business called Die Brüder Franz[3] in Bruchsal, a city not far from the Rhine in the state of Baden. Later, Karl Johan continued the business with the help of his four sons—Fritz, Karl Anton, William, and Joseph—as they became old enough to work. Fritz stayed to help in the store while Karl Anton, Joseph's father, studied chemistry in Vienna. Karl Anton made his own father very proud because he was the only offspring to have a university education.

In the late 1860s and early 1870s, an economic depression drove many Germans to immigrate to America. Karl Johan's oldest daughter, Theresia, and her family left first. In 1868, his son Fritz and his young wife Elise went to America. Karl Johan's next eldest daughter, Lina, remained home in Bruchsal.

When Fritz left, Lina's husband, Friedrich Hettrich, moved his office into the store to help. A young architect, he understood the materials used in the building industry. Several years later, Karl's younger sons, William and Josef, assisted their father, until William became sick in 1880 at the age of sixteen. Only twelve-year-old Josef was left to help his elderly father. It was difficult for Karl to manage the business.

When Karl Anton finished his education at the University of Vienna, he found an excellent job as a chemist at the Taussig Factory in Vienna. It produced very fine and well-known cosmetics. Karl Anton invented the first clear glycerin soap, which took the silver medal at the 1878 World Exposition in Paris.

---

[3] Franz Brothers

Figure 3—Karl Johan Franz, Joseph's Grandfather

Figure 4—Karl Anton Franz, Joseph's Father

After becoming well-established in his job, Karl Anton married his Austrian-born wife, Barbara, in 1874.

Figure 5—Barbara Tinauer Franz

Over the next four years, three of their eight children were born—Karl, Barbara (known as Wetty), and Ludwig. In 1880, filial duty finally drove Karl Anton to uproot his family and move home to work in the family business through this difficult time. Shortly after they settled in Bruchsal, daughter Maria was born. Two years later, on July 21, 1882, Bernhard Josef[4] made his howling entrance into Karl Anton's family. Karl Anton's and Barbara's fifth child was named for his Uncle Josef. Four years later, another daughter, Rosa, was born.

---

[4] In America, he changed his name to Joseph Franz.

**Figure 6—Two-year-old Josef Franz
and his brother, five-year-old Ludwig**

By that time, Barbara had exhausted her tolerance of the German rigidity and what she felt was an oppressive culture. A big-city person, Bruchsal was just too small for her. It wasn't a bad place with its charming pink palace, several fine churches, and lovely woods surrounding it, but she missed her family and friends with the music and laughter in Vienna. She was also physically and mentally worn out, resulting from raising so many youngsters while her husband worked in the family business.

After Rosa's birth, Regina Baroggio, a great-aunt, provided relief when she offered to take nine-year-old Wetty for a few years. Regina and her husband shared a big house in Ostringen that had seemed empty since their only daughter had married and moved away. Wetty also provided them with cheap household help. Regina was a kind woman. Though Wetty had to work hard, she got a lot of loving attention without having to compete with her siblings. Along with manners, she learned how to run an efficient home and cook some gourmet dishes. Best of all, she had her own room.

In 1886, two years before Amalia came to America, the Vienna Taussig Factory offered Karl Anton the position of factory manager. Barbara couldn't have been happier. Karl Anton's brother, William, now twenty, had completely recovered from his illness. His youngest brother, Josef, at sixteen, was quite capable of handling more responsibilities in the business. Recognizing an extraordinary opportunity had been offered to Karl Anton, his father wished him well in the new position. Thus, Karl's family packed their possessions and moved to Vienna. Wetty stayed in Germany until she was thirteen.

It was a great honor and joy for Karl Anton to receive such advancement in his career, but he was saddened to leave his aging mother and father as well as the rest of his family. It was the last time he saw his mother, as she died in 1888. That same year William died, leaving a wife and two small children.

In Vienna, in 1888 a son, Edward, was born. Two years later, a daughter, Alma, completed his family.

*   *   *

The move to Vienna that Amalia described that afternoon was only a vague childhood memory for Joseph. Amalia was the only relative in America that Joseph knew at all. Moreover, he barely remembered meeting her in Bruchsal. He was only six when Fritz Hettrich brought Amalia, age twenty-one, and their sisters, Maria and Rosa, to America after their parents died. Nonetheless, Amalia's kindness made Joseph feel at ease. They talked about the family still living in Europe.

Joseph learned that Amalia's husband, Karl Fetzer, was a pastry chef in a restaurant. She had met him in New York City, but they had moved to Chicago for several years after they were married. Their two young daughters, six-year-old Rose Marie and four-year-old Caroline, had been born in Chicago.

When Karl got home, all of them took the 3rd Avenue El to 86th Street for Joseph to meet Uncle Fritz Franz, Aunt Elise, and their daughter, Valeria. Fritz looked similar to Joseph's grandfather, but he was much heavier. Fritz was master confectionery chef at the Waldorf Astoria Hotel on 5th Avenue and 34th Street, where the Empire State Building now stands.

Because the Fetzer children were very young, Joseph was supposed to stay with Friz, Elise, and Valeria until his trunk arrived. This was the first time Joseph had met Valeria, who was about the same age as him. He was surprised she was still in school and thought the American school system must not be as good as those in Europe because it took so much longer to graduate. Joseph occasionally took Valeria on dates when he lived in New York City. She was good company, though not very athletic. She wanted to get married, but Joseph wasn't ready. He wanted to get established with a real career before settling down with family responsibilities. Her parents were a little disappointed that they never got together.

After dinner that first day, Fritz took Joseph to meet the Bendlers. Mrs. Bendler was Amalia's sister, Marie. When they returned to Fritz's place, Joseph's head was spinning with names and faces. As his weary eyes closed, he supposed all of it would be sorted one day.

The following Sunday, Fritz, the Fetzers, and Joseph took the 14th Street ferry across the North River to visit the Comisar family in Hoboken, New Jersey. These cousins were related through Theresia Franz. She and her husband had come to America some time before her brother, Fritz. Her husband was dead, so she was living with her son Alex and his family. Joseph vaguely remembered his father being angry with Theresia's husband. He had borrowed money from him to come to America, but he had never returned the money. As far as Joseph was concerned, that was ancient history. He was happy to meet these people because a couple boys were around his own age. This was the first of many family gatherings he would share in the coming years.

Once his trunk arrived the following week, Joseph rented a room in the home of Fritz and Luise Hettrich, Amalia's brother and his wife. After work, Karl hired a horse and wagon. He picked up Joseph and his trunk to take them to the Bronx to his new home at 633 St. Ann's Avenue.[5] Luise Hettrich was a truly kind, generous person. Joseph would learn to love her as his American Mother, even though he always called her "Aunt" and Fritz "Uncle." After all,

---

[5] The home was located in the Melrose District.

they were aunt and uncle to little Caroline and Rose. Their daughter, Hermine, was already married, so Joseph had his own room. He promised it would stay until he got established. He paid $2.50 a week for his room and board.

Figure 7—Fritz Hetrich and his wife, Luise Laier Hetrich

Chapter Two

Settling in New York

Joseph spent the first days with the Hettrich family, mainly telling them about his life in Europe. Fritz knew very little about Joseph's family. During the first few days, Luise spent many hours listening to Joseph recount his early years, starting with fact that his family had called him Bernhard Josef.

*   *   *

Young Bernhard Josef's formal education began in Vienna at the age of five. He learned quickly and easily, but he sometimes got into trouble because he was bored. He had plenty of encouragement from his father when he did well, and he was rigidly disciplined when his progress wasn't so positive. He began asserting his individuality when he insisted on being called "Josef" Franz in school, a name he preferred to "Bernhard," which he felt was too pretentious.

In those days, there were eight years of compulsory education, including five years of primary school and three years of middle school. Monday through Friday, a school day was eight hours long from 8:00 AM to 4:00 PM. On Saturdays, the students went until noon. On many Saturday afternoons, teachers took their pupils on interesting field trips to a factory, park, or nature preserve. These were opportunities for students to learn firsthand how things were made and about plants, trees, birds, and animals. In one or more of the many museums, they studied various forms of art. When the weather was warmer, they often attended free band concerts in the parks. There wasn't any physical education in school, but young Josef and his friends flexed their muscles after school about twice a week at a nearby public gymnasium.

On Sundays, family members took long walks together all year long. They walked in the various parks, woods, and fields outside the city. They also just went around to different areas within the city. Josef and his brothers went swimming together in the summer and ice-skating on frozen ponds in the parks or on the Danube River in the winter. Both Josef and Ludwig pursued these two sports with lifelong enthusiasm.

Josef was seven when the Taussig Factory opened a branch in Switzerland. His father was appointed its first manager. For two miserable years, the family lived in Wintertur. The Swiss were very conservative and prejudiced against all foreigners. Kids chased the Franz children home from school, yelling what seemed to be derisive threats all the way. Though the family was living in the German-speaking part of the country, it was hard to understand the Switzerdeutsch dialect.

One day, Josef's mother came home without any food because she didn't understand the shopkeepers. It seemed to be a completely different language. For example, *nähen* is the German word for "sewing." In Switzerdeutch, it is *büssen*. In another instance, when Josef's mother was once walking with baby Alma, several people asked, "Ist das ein Kind?"

Kind is the German word for "child," so she thought they must be very stupid not to realize that Alma was a child, not a doll. It wasn't until one person asked "Ist das ein Kind oder ein Bub?" that she realized *Kind* meant "girl." The family survived the ordeal of living in Switzerland for two years, but they were greatly relieved when Karl Anton was reassigned to Austria.

Wetty happily rejoined the family when they were settled again. Of all his siblings, she was Josef's favorite. Being five years older, Wetty became his surrogate mother and constant companion. She was the one who shared his confidences, listened to his dreams, and consoled his disappointments.

Even though language was no longer an issue in Vienna, there were other problems. The city was expanding rapidly, and streets were built around land excavated for new construction. Those spaces were irresistible battlegrounds for many street gangs that roamed the city. Gangs often spilled into residential neighborhoods and hassled kids en route to school. They demanded money or anything of value and always tried to recruit new members.

Josef was an ardent pacifist all of his life and hated violence of any kind. He stalwartly refused to join any gang, even when threatened with bodily harm. Though still a youngster, he showed signs of his potential by finding creative ways to solve his problems. With logic and reason, he earned respect and protection from all of the gangs, regardless of who was fighting. He became their Red Cross! He had a small, wooden medical box that he carried over his shoulder. The box was filled with cloth for bandages; thread and needle for sewing cuts and gashes; iodoform for cleansing wounds and alleviating pain; paregoric for relieving the pain of a broken tooth and pain in the jaw; and a salve for treating bruises. He treated anyone who was hurt. He even had one old man, who had very bad teeth, as a steady patient. Josef regularly treated his aching gums with paregoric.

For vacations or school holidays, Josef sometimes worked in his grandfather's store in Bruchsal. He wasn't afraid to travel to Germany by himself on the train. In fact, he dearly loved the independence he experienced and the affinity he felt with the great coal-eating beast as it rolled along the iron track. On the train, he could dream of marvelous adventures and imaginary inventions that would save the world or at least that part of the world he knew. After all, Germany and Austria were at the forefront of new scientific discoveries. Why couldn't he be a part of it, too?

Reality always set in when he reached Bruchsal. First, he was called Bernhard, so there wouldn't be any confusion between his Uncle Josef and him. To do his job, he mainly stayed outside in the brickyard, stacking supplies of bricks and hollow concrete blocks. Bricks and blocks were some of the many building materials that his grandfather sold. It was hard work for the youth, but it was also good outdoor exercise. He enjoyed being outside, away from the stuffy classrooms in the city. The exercise helped build his muscles. There wasn't any pressure to compete in Bruchsal, and his grandfather always seemed grateful for his help.

Back home, among other subjects, English, French, and art were taught in middle school. Art was his favorite subject, and his teachers praised his excellent work. He kept some of his work from those years, just to remind himself of his artistic capabilities. One time, the art class was learning to draw a still life. The subject was a bowl of fruit containing a banana. It was the first time the students in 1895 had ever seen a banana.[6] At the end of the lesson, the teacher cut the banana into small pieces so that everyone could have a taste. Josef never forgot that delicious flavor.

As a special treat, Josef's father occasionally took him and other family members to an opera or concert. The parks still resounded with the music of free concerts and other entertainments in the summertime. Museums offered changing exhibitions of paintings and sculpture. All of the public office buildings, theatres, and palaces were examples of elegant architectural styles. The arts became as much a part of Josef's life as his academic studies.

---

[6] A banana was considered a luxury.

Because he was also interested in science and the new technology of electricity, he went on to the Wien Techniche Gewerbe Schule[7] when he finished his compulsory education. He studied subjects such as French, physics, mathematics, and engineering. He also took classes in German literature, ancient history, and political science. His formal education ended in August 1897 when he received a degree in electrical engineering. To celebrate his achievement, his father took the family out for dinner in one of the fancier restaurants in the city. He even ordered a bottle of wine and served some to Josef. Josef felt proud, and he was glad his hard work was appreciated.

He had done a lot of thinking about what he would do after finishing school. Because his father was German and he had been born in Germany, Josef would face enlistment in the German military service at the age of sixteen. Under the domineering decrees of Chancellor Otto von Bismarck[8], he'd have to serve three years of active service and four years in the reserve. Having turned fifteen in July, he knew he'd have to register for the army in a few months. The very idea was abhorrent and deeply frightened him.

He thought about the great things happening in the United States. With the increased industrialization, he thought there would surely be a need for a bright, young electrical engineer. America couldn't be such a bad place. After all, his Uncle Fritz, Aunt Theresia, and several cousins had already left for America. Moreover, they seemed to be well-established. They at least never returned to Germany. After some convincing, his parents reluctantly agreed to let him try his luck in New York City, but he had to promise to never forget his family or from where he had come. If he was unable to make the adjustment, he'd always be welcome at home.

To enter the United States as an immigrant, one needed to have a passport and at least thirty dollars (US). One also had to be in good health. When his father came home with the final requirement on October 4, 1897, Josef was ecstatic. He now knew he was really on his way. His small trunk was sent on ahead of him.

Two days later, after saying good-bye to all his friends and relatives, Josef headed to Vienna's North Station with his father, mother, Marie, Rosa, Eddy, and Alma. All of them came along to see him off. Wetty was away in a hotel management school near Munich. Karl and Ludwig were both serving their

---

[7] Vienna School of Technical Industry

[8] A Prussian Prince, appointed by Emperor William II

military service. His sisters, mother, and Eddie gave him hugs and kisses. His father gave him a large bear hug and patted him on the head.

He then boarded the train, which promptly left at 6:00 PM. As his family diminished into the darkness, Josef wondered when or if he would ever see them again. The doubts soon vanished into the falling snow, and he sank into a sound sleep.

Sleeping almost all of the way to Leipzig, he awoke when a boisterous group of students boarded at Bad Lausick. They were en route to the Leipzig University. By the time they reached Leipzig, where he had to change trains, it had stopped snowing. He checked in at the Bahnhof[9] Hotel because his train to Bremen did not leave until early the next morning.

Leipzig is known for its university and being a center of commerce. At the large Oktober Markt,[10] which wasn't far from the station, Josef bought a lot to eat from food vendors and looked at various items for sale from foreign and domestic origins. The excitement of departure and the anxiety of the previous week finally caught up with him, so he returned to the hotel and fell into bed without undressing at 7:00 PM. He slept soundly until his alarm clock woke him at 3:45 AM. He washed up a bit, but, because his train left at 4:30 AM, he did not bother with breakfast. He knew he could get food aboard the train.

On the train, Josef slept until the food vender came down the aisle with his small cart. After coffee and *brotchen*[11], Josef surveyed the passing landscape that changed from rolling hills to plains and from agriculture to industrial cities. He wasn't a stranger to train travel, having taken several trips with his family or going alone to spend vacations with his grandfather. Anything that rolled, hoisted, or, later, flew was a fascination for him.

He arrived in Bremen on October 8 at 4:30 PM. He went to the steamship agent's office to pick up his ticket. However, he found it was in his father's name. He could correct the name on the ticket, but he wasn't able to change it on the passenger list because it had already been printed. Luckily, the office was near the Bahnhof. It was another short train ride to Bremerhaven, where he boarded the *Königen Luise*.[12]

---

[9] Railroad Station
[10] October Market
[11] small rolls of bread
[12] Queen Louise

A telegram from his father, dated from October 6, waited for him. It said it was snowing in Vienna and all the family was well. He then wished Josef a good voyage. Knowing his family was really seeing him off in good spirits, Josef was happy. At 12:45 AM on October 9, 1897, the ship left her moorings and slipped out of the harbor, taking Josef to a new life in America.

At the beginning of Joseph's first sea voyage, he watched all of the crew's activity. It fascinated him as they docked in England to take on more passengers. From then on, they were on the open ocean. All activity was strictly aboard the ship. He spent a lot of his time on the deck. The sea was calm, and the only sounds were from the hum of the ship's engines and the soft voices of passengers and crew as they went about their tasks to keep the vessel running smoothly. Mealtimes were always a pleasure when he consumed his favorite meats and potatoes. He sometimes had seafood soup with his meals, and he generally had wine or beer as a beverage. In mid-morning and afternoon, he ate pastries or sweet breads and coffee. Josef was particularly interested in the infirmary because he could build on his limited medical experience. Seasickness was the most common complaint. All would eventually recover. Best of all, he discovered he could get coffee and a roll there, even in the middle of the night

The last two days on the ship seemed endless. He could only look at so much ocean. Even though several ships were seen, they were soon out of sight. The crew's routine varied little from one day to the next. As the days dragged on, Josef woke earlier in the morning. He stared at the ceiling in his cabin, wondering what he would face in the new land. Each day, his thoughts grew more disturbing.

"What if my cousins are no longer in the city?"

"What if I can't find my Uncle Fritz?"

"Will I be able to read the signs?"

"If I can't find my relatives, how will I find a place to stay?"

"Will my trunk be there and be safe until I can pick it up?"

"How will I communicate in German or with my very little English? *Mein Name ist* Joseph!"

On Friday, October 22, his anxieties melted. Excitement overtook him when the ship finally sailed past the Sandy Hook Lightboat and went into New York

Harbor. He was up at 3:00 AM that day. He quickly drank coffee and hurriedly ate a roll. He packed his last belongings, and he was on the deck with his small bag at 4:00 AM, ready to leave the ship. Punctuality was a habit he developed to a fault.

In that predawn hour, the beckoning lights of the Statue of Liberty could be seen all the way from the Narrows. People were pointing and shouting.

"There it is!"

"Look, it's the statue!"

One woman got down on her knees to express her thanks in prayer. The emotional thrill, generated by the reactions of his fellow passengers, quite surprised Joseph. The closer to Lady Liberty they came, the stronger it grew. The feeling of warmth and welcome would sustain him many times throughout his life, even in the darkest hours.

* * *

The retelling of Joseph's story to Luise brought back memories of her own arrival in the country. The two of them developed a longstanding friendly bond.

After settling into his new surroundings with the Hettrichs, he realized he needed to immediately find employment. He didn't want to be dependent on anyone. Plus, his thirty dollars wouldn't last very long. The day after he arrived, Fritz and Karl took him to a penny bank to possibly get an odd job. He managed to repair a telegraph extension. For four hours of work, he received $1.25, with which, he bought lunch and a beer.[13] Plus, thirty-five cents were left over!

During the next month, he worked in several neighboring apartment houses, polishing the brass doorplates and knobs and other brass fixtures. For this, he received between ten and twenty cents a job. His first real job was in one of the tenements on Mott Street in Lower Manhattan. He was hired to paint the hall ceilings for one dollar a week. He must have done a good job because he got two dollars for the second week. After that, he worked for Austeicher's Construction Company, helping install electric wiring for three dollars a week. He also worked various odd jobs. Ultimately, the pay he

---

[13] Back then, there was not an age limit for beer or wine.

received did not cover his expenses. As expected, the thirty dollars was gone quickly. He had to borrow another ten from Luise.

Six months after his arrival, at the end of March 1898, he finally secured an apprenticeship in mechanical engineering with Francis Keil and Son, a tool and dye company. An apprentice working at the drop hammer had lost a hand when the hammer fell too soon. Joseph was hired to take his place.

Figure 8—Francis Keil and Son, Tool and Dye Company;
Joseph (inset) at drop hammer

In the year and nine months he worked with the firm, he learned to operate all kinds of machines and acquired many mechanical skills. The apprentices took turns on shaping and stamping machines, including work forging iron, brazing brass, burnishing, polishing, and plating other metals. The shop made all kinds of metal products, including bells, buzzers, gears, and plumbing supplies. He learned to operate rolling, planing, cutting, and forming machines. Working conditions were bad in the basement shop.

There weren't any safety devices on the machinery. Some of the workers were hurt badly.[14] Thankfully, Joseph's studies in Austria had provided him with a basic understanding of the various machines. He also listened intently to instructions. He managed to escape any serious injury, though, despite his efforts in safety, he did get a few minor scrapes and cuts.

When he didn't have to be at the tool and dye company all day, he did odd jobs, for example, wiring apartment house bells. With the extra money, he was able to repay his debt to Luise and open a savings account at the dollar bank. He sometimes worked for a friend of Fritz and Luise, Mr. W. J. Fitzgerald, who owned a delicatessen on Columbus Avenue. In that store, he saw bananas for sale. He remembered the delicious taste from his art class experience years before and recklessly bought a whole bunch. On his way home, he consumed every one. Even though he got sick that night, he still liked the taste. However, from then on, he ate them one at a time.

Joseph's father taught him to take care of his money. Growing up, he had to keep an account of every penny he got and where it was spent or saved. This habit had become so much a part of his routine that he continued the practice of keeping exact accounts when he came to America. For example, as his income increased, his rent was raised. In 1898, his lodging was raised from $3.50 to $4.00 a week. His pay ranged from $4.75 to $5.50 for six days of work, but he somehow managed to live within his salary. Transportation was five cents, and he only ate the two meals a day that Luise provided. After work, he often went to the Fetzer's home or Luise's mother's place for dinner and family news. By the end of that year, his bank account was up to $45.30.

Even though work consumed most of his time, he did enjoy intervals of relaxation and fun. In the fall and spring, he'd play football[15] with fellow workers during his lunch breaks. An evening's entertainment was usually a walk. Sometimes, he was alone. Other times, he went with friends. In the summer, he took jaunts to Oak Point Park.[16] He also sketched in Central Park.

The Bendlers lived near Fritz and Luise in the Bronx. These relatives would often exchange visits. Fritz occasionally took Joseph with him for a beer at a

---

[14] Laws concerning the work of people between the ages of sixteen to eighteen were not enacted until 1941.

[15] It was called soccer in America.

[16] The park is located on the East River.

local saloon. On Friday or Saturday evenings, he'd join one or more friends from work for a beer, but rarely more than one.

Fritz and Luise recommended that Joseph start English classes in Evening School 62 on 157th Street. Even though he had studied some English in school in Vienna, he didn't have any chance to practice it in Europe. For two years, he attended English classes in New York City each fall and winter. He learned quickly. Soon, he could speak fluently with very little accent and read newspapers, magazines, and books. Writing came more slowly. Spelling was always difficult. Then again, he wasn't ever a very good speller, even in German. Music was still an essential part of Joseph's existence. Even though he didn't have a very good voice, he enjoyed singing with groups from the English school or alone. It was a very special treat when he was taken to a concert or the Metropolitan Opera. It didn't matter that they sat in the top balcony.

On his only days off, Joseph's interest in electricity and electrical devices increased. Sunday was the only day when he had enough time to experiment. He built an induction machine and several other electrical apparatuses. He'd entertain relatives and friends with them. The lightening-like flashes from his Leyden jars really scared little Rose and Caroline, but it was fun most of the time.

The Laiers, Luise's family, lived on East 32nd Street. On many Sundays, other relatives and friends would be there. Mrs. Laier, who seemed to be quite an old lady, loved to entertain. She played the piano very well. Their apartment was bigger and more centrally located than Luise's place, making it easier to gather there. Joseph could only take about an hour of family gossip. Excusing himself early, he met his friends. They'd go to a park or, if the weather was bad, someone's home. He also enjoyed his frequent trips to Hoboken, where he played ball games or just talked with the Comisar boys.

Holidays were generally large family gatherings at either the Fetzer's or Laiers' home. Joseph was amazed how many relatives had migrated to the New York area and were raising families in this new land. As the children grew older and married, they started new families. The Bendlers had two. The Fetzers had two, and the Comisars now had five. There was always plenty of food with wine and beer because everyone brought something for the occasion.

Joseph's first real American holiday was on September 29 and 30, 1899, a celebration to honor the Spanish-American War hero, Admiral George Dewey. This was the first time that Joseph had really been made aware of American politics or current history. He had been too busy with work and relatives to have much time left over for news that did not directly concern him. This was something special. While reading about this celebration he learned the United States had waged a war outside its own land for the first time. Admiral Dewey had been successful in defeating the Spanish fleet in the Philippines so America acquired that country, Guam, Cuba, and Puerto Rico as colonies.

The city had built a beautiful victory arch and colonnade on Fifth Avenue at 23rd Street for a big parade that took place on September 30. Joseph didn't get to see the parade because he was working, but he did see the spectacular display of fireworks the night before.

Two months later, Joseph had his next introduction to American celebrations at Thanksgiving. The Fetzers invited him for a traditional turkey dinner, which he found very tasty. On that occasion, Rose Marie and Caroline told him the story of the pilgrims and the Indians and how everyone was thankful they had survived their first year in America. Joseph thought about his own first year in America and decided he was also thankful to have survived. Other holidays, except Christmas, went by unceremoniously. They were simply days when he didn't work.

Christmas was generally celebrated on the evening of December 24. Someone would have a Christmas tree that everyone would decorate. Several baked goods, cookies, cakes, candy, and *glühwein*[17] were served, and a few gifts were exchanged. No one had a lot of money, so no one felt obligated to give something to everyone. Joseph always got something from Luise, like gloves, socks, or other items of clothing. He always gave her something in return. He usually spent Christmas Day visiting various friends and relatives.

For the eager youth, time still seemed to pass slowly in those early years. He spent many lonely hours adjusting to a new country, language, and people. Joseph wrote many postcards to friends and family at home. He discovered he could get a postage stamp-sized photo made on the Bowery for twenty-five cents, so he sometimes managed a letter with an enclosed photo. In return, he received many cards and letters, some with photos. His mother and sisters were the most faithful. To alleviate his boredom, he began saving the stamps from the letters he received. This started his interest in collecting stamps and coins, a hobby that lasted for many years. The family continued writing even after Joseph became so absorbed with work that he didn't have much time to respond as frequently.

---

[17] Hot, spiced wine

In one of the letters from home, he learned his brother Ludwig finished his education as a chemist, like their father, when he got out of the army. While working in Hamburg and Berlin, Ludwig wrote to Joseph a lot. He eventually returned to Vienna, where their father could get him a job at the Taussig Factory.

The biggest surprise was a telegraph cable from his eldest brother, Karl, in December 1898. He said he was coming to New York City the following June. He was out of the army and couldn't find steady work in Austria. Luise said he was welcome to stay with her and share Joseph's room.

When Karl finally appeared on June 14, 1899, Joseph was unable to get off work to meet him. Predictably, the always kind and helpful Luise was there to welcome him. Surprisingly, he wasn't alone. Karl was twenty-five and had married just a couple weeks before leaving Vienna. With his recent bride, Mary, he was anxious to settle quickly in the new land. He felt they would be better off in New Jersey, which was less populated. Luise helped them locate a reasonable hotel until they could find an apartment. In a couple days, they found a place in Hoboken, not far from the Comisars. Karl was an accomplished violinist, but he knew he couldn't support a family on music alone. As a temporary solution, Karl quickly found a job in a German bakery. Soon, their daughter, Theresa, was born in 1900. She would be their only child.

Joseph was thinking about a new job as well and the possibilities open to him. At the turn of the twentieth century, several independent power companies were in New York City. Some companies in older parts of the city had overhead systems, which looked like spaghetti factories because of all the wires strung in complete disorder. Other companies in newer sections in the city supplied electric power via an underground system. They had buried the electric cables and telephone wires under the streets. Over many years, Manhattan and much of the rest of the city converted to underground wiring.

In November 1899, Joseph saw an ad in the newspaper for someone who could wipe lead joints. He applied for the job, even though he had very little professional electrical experience. When he went to the office, the boss, Mr. Penden, asked if he had ever wiped joints before. Although Joseph hadn't, he had studied and read how it was done. He knew he could learn how, so he said yes. He was told to report for work at the North Service Electric Company of New York the next morning. Wiping joints meant splicing electric cable. The ends of two pieces of cable were cut on an angle. The lead covering and insulation were peeled back so the two ends of copper wires could touch. The individual wires were then connected. The lead cover and insulation were pulled back over the joint as far as possible. The whole splice was finally covered with hot molten lead.

En route home, Joseph picked up some lead, a length of cable, and other supplies. He spent the night in the basement, practicing the art of wiping lead on the joints of cables. When he showed up on the job, he was an expert lineman. That was his first professional work in the electrical field. His superiors highly praised him, and he was justly proud of his accomplishment.

While working as a lineman, he was part of the crew that installed electric cables in Upper Manhattan from 127th Street to 170th Street. Because of his splicing skill, he was sometimes called upon to fix a broken line elsewhere in Manhattan. His ability was soon recognized. and he was assigned to work in several powerhouses, connecting the new lines, installing transformers, and maintaining the various kinds of equipment.

Joe Klein was in that first crew. Joe was about the same age as Joseph, and the two of them became good friends. He and Joseph palled around together on Friday and Saturday nights. They mostly just walked and talked because neither one had much money. Joe taught Joseph a lot about the city and America. On Sundays, they'd play soccer in a park or go boating. W. J. Fitzgerald, the delicatessen man, occasionally joined them, always supplying a delicious picnic from his shop. A favorite spot in the Bronx was Oak Point Park. The park bordered the Bronx River on the south, where the river empties into the East River.

With steady employment, Joseph decided he earned enough money to afford a place of his own. It wasn't exactly his own, but it was a place he didn't have to share with friends or relatives. For most of the next year, he stayed in a boardinghouse on St. Ann's Avenue, not far from Luise.

Joseph worked for several bosses until May 1901. One, a Mr. Wilson, gave him good references several times. In early 1901, he introduced him to a completely new line of work. Mr. Wilson arranged an interview with Adolph Zuckor,[18] Joseph's first fleeting brush with the entertainment industry. It didn't lead to immediate employment, but it was a prophetic glimpse into his

---

[18] Born in Hungary in 1873, he came to the United States at fifteen and worked in the fur business. He later bought the fur company, which he sold to buy his first penny arcade. These led to nickelodeons. In 1912, he bought the rights for a French film, *Queen Elizabeth*, with Sarah Bernhardt, which became a big success. He eventually became the founder of Paramount Pictures, where he remained chairman of the board until his death at 103 in 1976. He was awarded an Oscar in 1948 for "his contribution to the industry."

future, which would bring more lucrative and personally gratifying associations with the arts.

About that time, he began noticing girls as more than just different from boys. Several friends had attractive sisters, but, whenever they were around, there weren't any opportunities to be alone with them. Society still frowned on dating without a chaperone. He confided to Joe Klein that sexual urges were beginning to be bothersome. Several of their more experienced friends suggested that Joseph and Joe join them the next time they visited the girls. There was a house in Brooklyn where the girls lived and worked. Joseph and his pal eagerly accepted the invitation a week later. Having discovered new pleasures, Joseph found his own way back there on other occasions. After such a visit, he always had a very good night's sleep.

In the winter, Joseph and his friends began ice-skating in Van Cortland Park and on other rinks in the area. Joseph had started skating as a kid in Vienna with his brothers. Their first skates had been the kind that clamped onto their shoes. During the winter of 1901, he bought his first pair of shoe skates with racing blades. Shortly after getting the new skates, he had a bad fall and injured his ankle. After a few days of rubbing it with witch hazel, he was skating again. This was one of his favorite sports throughout his life. A toboggan run in Van Cortland Park offered an alternative sport.

For relaxation from the weekly grind in the summer, there was swimming or boating in the East River off Ferry Point, the big park that now lies under the Whitestone Bridge. Joseph usually went with a friend or a relative. With a pitcher or bucket of beer from a local bar, they'd head for the water. However, Joseph once took a rowboat out by himself. He liked rowing out from shore, getting in the water, and just floating around on his back. This particular day, he must have floated for a very long time. When he looked up, expecting to get back in the boat, he noticed it was far away. Currents in the East River are very strong. The tide was going out, and the boat had gone with it. He was a good swimmer, but it was miraculous that he was able to catch up with the boat and still have the strength to pull himself into it. When he flopped in, he just collapsed. His whole life flashed in front of his eyes. He was sure he was dying. He laid there until well after dark before reclaiming enough strength to row back to shore. It was 2:00 AM before he fell into bed. He had thankfully survived his first hard lesson, and he never took a boat out alone after that.

Chapter Three

Becoming Self-reliant

Joseph was asked to join the electricians union in May 1901. He knew he was now accepted as a professional electrician because he was immediately hired as a troubleshooter for the Boston Emergency Service Company. Only nineteen, he was short with a slight build, so he still looked like a kid. Many of his fellow workers were lot older and bigger than he was. In order to look older and, hopefully, more important, he grew a mustache. On the train to Boston, he never doubted his abilities as far as the job was concerned. He did wonder if the mustache would really help win the respect he needed. It must have achieved the desired effect because it stayed with him most of rest of his life.

Figure 9—Joseph, age nineteen

Housing was booked for him on East Railroad Street in Dorchester.[19] The accommodations were perfectly adequate, but the neighborhood was mostly industrial. Noisy during the day, it was deserted at night. It wasn't conducive for evening walks or even finding a decent restaurant. In a couple months, he found another boardinghouse run by John and Berthe Boucke on Dorchester Avenue in South Boston. The boardinghouse was in a block of attached row houses. Each one looked exactly like its neighbor.

Joseph worked odd hours that week. Working the evening shift, he left around 5:00 PM, right after he moved into his room. When he returned home early the next morning, he couldn't find the house. He knew it was on Dorchester Avenue, but he had forgotten to take the new address. Because he was tired and all of the houses looked alike, a real panic overcame him. For what seemed like hours, he wandered up and down the street. About 8:00 AM, Mrs. Boucke came down the street. She was on her way to the store for groceries. She fortunately recognized her new boarder and laughed when he admitted his stupidity. She pointed to the house, which he had passed many times. He immediately sketched a picture of the house in his notebook and wrote down the address. Wherever he stayed from then on, he made sure he had the name, address, and phone number in his wallet.

The Bouckes were very kind to Joseph and treated him like a son. In return, whenever he had time, he was happy to help repair things around the house, shovel the snow, or rake the leaves. He would sometimes have a drink with Mr. Boucke or play a parlor game with Mrs. Boucke.

In the Boston area, most of the power lines were run overhead on poles. Joseph repaired electrical failures whenever and wherever they occurred in the contract location. Short circuits that came from overloading the power line caused many of the failures. This often caused a small fire in a transformer on one of the poles. Other failures were due to broken lines. He rarely got a decent sleep because much of the repair work began at 3:00 AM. When not on a call, he worked on transformers and other equipment in the North Point, Arlington, Central, or Cambridge repair stations. He was on call twenty-four hours a day. His days off changed constantly. At times, he couldn't sleep for several days because his schedule was so erratic.

To relax at odd hours of the day, Joseph was thankful for his artistic training. He drew pencil sketches or made charcoal renderings from life or photos.

---

[19] Dorchester is a community within the southern section of Boston.

This art included still lifes, animals, scenery, and even a portrait of Henry Wadsworth Longfellow. When all else failed, he'd visit the local girls to get a few hours of a sound sleep.

He didn't have any problems with the men around him on the job. Most were Irish immigrants. After work, Joseph had a drink with McCarty, McDonnell, Gill, Sam, or Jerry. Crews changed frequently, so Joseph never knew the men by more than one name. In the summer, they went swimming together. In the winter, they went ice-skating. He unwisely trusted everyone by trying to make new friends. When his friends overspent their paychecks, they asked to borrow money from him. They always promised to pay it back when they were paid, but, except for five dollars, no loan was ever repaid. He learned another hard lesson.

For a short time, he joined the Boston Brownies, a club of year-round swimmers who believed a swim in the icy, wintry waters was good for one's health. In his youthful enthusiasm, it sounded reasonable. However, when the president of the club died of a heart attack in December, Joseph resigned from the club.

Joseph soon knew Boston as well as he knew New York City because emergency calls came from all around the area. With increased pay, he was occasionally able to attend theatrical events at the Morrison or Kiel theaters. One highlight was when he saw *Faust* in the opera house. During the summer, he took advantage of the public concerts in the local parks.

International or domestic events did not affect Joseph very much until he heard about the assassination of President William McKinley in September 1901. It became a great personal tragedy to Joseph and his coworkers. President McKinley was shot shortly after his second inauguration.[20] Joseph knew President Abraham Lincoln had been killed, but that was history. European history was full of such political violence. Even at the end of the nineteenth and the beginning of the twentieth centuries, Europe's crowned heads were being eliminated. German and Italian unification, along with Austro-Hungarian Empire's expanding boundaries, had engulfed principalities and dukedoms. Joseph believed such atrocities could not happen in America. The night the president was shot, he and a coworker, Charley Massel, were drinking in a pub. They were not alone. The assassination stunned the

---

[20] McKinley was attending a reception in Buffalo at the music hall when Leon Czolbosz, a young anarchist, shot him. He died two days later.

whole country. In the next week, Joseph diligently read the newspaper accounts of the assassin's capture and inauguration of the new president, Vice President Theodore Roosevelt. Many buildings in downtown Boston were draped in red, white, and blue bunting. After the president's death, flags flew at half-mast. The week before McKinley's funeral, as a memorial tribute within the company, the boss, Mr. Burns, brought schnapps and beer to the office at lunchtime. On the day of the funeral, they left work at noon.

Joseph continued working for the Boston Emergency Company. However, by November, the erratic work schedule was having an adverse effect on him. He had trouble adjusting his sleeping time. On many days, he barely had four hours of sleep, and he took that time in small naps. He was finally so disgusted with the situation that he looked for work with another company.

When the Brighton[21] Emergency Service hired Joseph, he took a brief weekend to see his friends and relatives in New York City. Joseph's family in Europe was still writing faithfully. Most importantly, his oldest sister, Barbara (Wetty), had married Karl Gumpinger. Karl and Joseph were good friends and frequently corresponded. On the train to Boston, Joseph wrote to congratulate his sister and warmly welcomed Karl into the family. He arrived refreshed and energized to begin work on Monday morning.

Most of the calls at this new job were related to the trolley lines. He didn't have anymore night calls, which made his life more normal. Toward the end of December, he was promoted to line foreman for a crew that was wiring the Boston transit system. They worked in several locations around Boston. They sometimes strung steel support wire between the parallel iron poles on either side of the tracks. The poles were easy to climb because they were constructed of two solid U-shaped iron rails. Iron straps riveted onto both sides of the solid pieces, like large Xs, held the rails together. The Xs grew smaller as they got to the top, making each pole a small tower. A wooden tower, mounted on a flatbed rail car, gave access to the support cable where they attached a ceramic insulator over the middle of the track. Embedded on the bottom of the insulator was a clip to hold the high-voltage copper wire that would power the trains. In other areas, they strung the bare copper wires, attaching them to the insulator clips on the steel support wires. In between or when the weather was too inclement, they worked on repairs in the car barns.

---

[21] Brighton is a northwestern community of Boston. It is located next to the town of Newton.

One cold day in February 1902, Joseph was working on a siding curve of elevated track near a car barn. The flatbed car holding a spool of wire was on the center track. He was on that car, threading the number five steel wire into the eye of a wooden strainer.[22] The boss, John Gillpatrick, was the driver of the flatbed cars. He was about a half-block down the track toward the car barn when the wire hit a snag. The spool car suddenly lurched forward and ran into the construction tower car. The jolt threw Joseph to the ground. He landed feet first, but he fell back against the edge of the car, injuring his back and knocking him unconscious. When he came to, he had a terrific pain in both legs. After being taken to a hospital, he stayed for a week-and-a-half. The company's insurance paid for all the medical bills, but they did not keep him on the payroll during his recuperation. Frugality and savings helped pay his other expenses.

Joseph's right foot, the same one he had injured in New York City while ice-skating, was slow in returning to normal. Mr. and Mrs. Boucke kindly tended to his needs so he could recuperate in relative comfort. As soon as he was able, he began walking. Later, he regularly went to the local gym. On April 1, he felt secure enough to look for work again.

The Lord Electric Company hired him to string wire and handle repairs in Boston. After a couple months, Joseph asked to be transferred out of the city to gain experience stringing long-distance lines. The company complied with his request by transferring him to the Haverhill office. Until September, he was part of a crew digging holes for poles, setting poles, cutting the cross arm supports, and mounting the pegs for insulators. He climbed poles to attach the arms, set the glass insulators, strung the insulated copper wire, and connected the lines to the appropriate terminals. At that time, poles were set only 100 feet apart because the copper wire was purportedly not strong enough to span a greater distance. His crew put in lines to the north as far as Salem, New Hampshire; as far to the west as Worcester, Massachusetts; and all of the points in between.

It was hard work, and he didn't have enough time for relaxation. However, he learned a great deal, which would serve him well on later assignments. Occasionally after work, Joseph would have a drink with one or more of his fellow crewmembers in the nearest town with a bar. Two of the crewmembers

---

[22] Strainers kept the power lines separated so they did not arc over if struck by lightning.

were a father and his teenage son. One night after work, the father discovered his son masturbating behind a bush.

Furious, the father yanked his son out from behind the bush, yelling "No son of mine is gonna be a goddamn fairy!" He promptly took the boy to the local whorehouse to initiate him in the ways of manhood.

Sex was one way to relieve the tension and physical exhaustion of the job. Every few weeks on a Saturday afternoon, Joseph went to a shore town's brothel with a friend. He also went into Boston by himself. During the summer, after a relaxing Saturday night, he went swimming on Sunday. He also visited friends or went to the theatre.

Now that he was making good wages, Joseph bought a few pieces of jewelry for himself, including a pocket watch and a couple of gold rings. He was extremely pleased with his treasures because they represented a measure of personal success. On one of his forays into Boston, he carelessly left the jewelry as well as his bankbook in a small traveling bag at the cheap hotel where he was staying. Someone broke into his room and stole the bag with everything in it. The bankbook was eventually returned to the Boston Bank. They sent it back to Luise Hettrich's address, but everything else was gone. He had learned another hard lesson. He didn't waste anymore money on personal jewelry.

Joseph's first venture into Western Massachusetts was on September 29, 1902. The Fred T. Ley Company of Pittsfield hired him to take charge of the electrification of the Berkshire Street Railway from Great Barrington to South Lee. Another crew was working on the section from South Lee to Pittsfield, the Headquarters of the northern section.

For most people, transportation between the towns was still either by horse or the New York, New Haven, and Hartford Railroad. The latter only ran twice a day between New York City and Pittsfield, Massachusetts. It traveled twice in the opposite direction. It was not convenient for those who worked in a neighboring town. The Berkshire Street Railway served the needs of the local people. The northern section had already been finished from Pittsfield to Williamstown.

Headquarters for the southern section were in Housatonic, a lively manufacturing town like its southern neighbor, Great Barrington. The trolley was to run north through Stockbridge and east to South Lee. After checking in at the office in Housatonic, Joseph had to find a place to live. Because he was to work in all of the above towns, he found a room in a nice boardinghouse run by Mr. and Mrs. Chapin in Glendale. Glendale is in the Stockbridge Township, about halfway between Housatonic and South Lee. (He did not know he would eventually live in Stockbridge.)

Stockbridge, like its northern neighbor, Lenox, was where wealthy urbanites from New York City and Boston had built their summer cottages. Those huge estates dotted the landscape with Italianate tiled roofs, Tudor turrets, and Tiffany windows. Gentlemen farmers enhanced their views of verdant hills and sparkling waters with fields of domestic farm animals, like cows or sheep, in the foreground. The area seemed to be the epitome of picturesque tranquility.

In most of the towns, the trolley ran down the main street. In Stockbridge, the citizens voted to keep their town in its pristine condition, free from the noise and clutter of an unsightly mode of transportation. Ironically, the man who invented the trolley system, Stephen Field, lived in Stockbridge. He and other members of his famous family[23] were some of the strongest opponents to the trolley being on Main Street.

As a result, from Housatonic through Glendale, the tracks ran on the opposite side of the Housatonic River from the railroad. In Glendale, it crossed the river, running parallel to the rail line, crossing the river again just north of the Stockbridge railroad station. It traversed South Street, continuing down what is Park Street today. Running east to the Housatonic River, the trolley crossed it on a steel truss span bridge and again paralleled the train tracks. At the east end of town, it crossed the river on another bridge and followed the main road[24] all the way to Main Street in South Lee.(See map, figure 10). Or whatever the number is of the trolley map.

---

[23] Cyrus W. Field laid the first Atlantic cable. David Dudley Field was a minister and "Father of New York Penal Code." Stephan J. Field was the last Supreme Court Justice appointed by Abraham Lincoln.

[24] Route 183

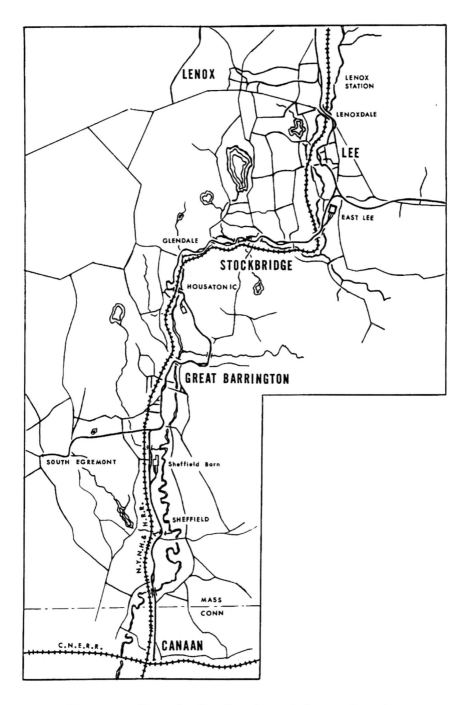

Figure 10—Map of trolley line through Stockbridge from "Berkshire Street Railway"

Tracks along the entire route were already in place. Most of the wooden poles had been set with perpendicular, galvanized iron hangers and guy wires attached to support the steel cables. The guy wires, attached to the top, ran to the bottom of each pole, grounding it in case of a lightning strike.

Joseph's crew strung the support system and copper wire to power the trolley cars. A steel cable was strung from the single pole. It went over to the center of the track to an eye bracket fixed vertically at the end of the iron hanger. A porcelain insulator was attached to the cable. Embedded on the bottom of the insulator was the clip to hold the high-voltage copper wire. To get to the support cable and insulator over the track, a wooden tower was mounted on a flatbed railcar.

One of the problems they encountered was due to vibrations or ice. The porcelain insulator over the tracks would break down between the two lines. It would fold, allowing the copper wire to touch the steel support wire and causing a partial grounding of the lines. Electricity would travel from the feeder line, over the support cable, back to the pole, and through the ground to the steel rails, thus creating a limited short. On several occasions, a horse stepped into the path of the ground line. The horse, wearing iron horseshoes, became a better conductor than the ground, especially soft, moist ground. The horse would be killed. The electricity wasn't strong enough to kill a person with ordinary leather shoes or boots. Joseph solved that problem by adding two strain insulators[25] on the steel cables on either side of the center feeder insulator. The bracket he created made it impossible for the copper wire to fold onto the support cable.

---

[25] A strain insulator is a porcelain object with two perpendicular holes, allowing the cable to interlock at two different angles.

Figure 11—Trolley line in Glendale showing power pole and line support

    Due to the intensity of the work schedule, Joseph only had a few events to attend during his leisure time over that fall and winter. Music was enjoyed only when he and several other crewmembers got together to sing. They once took a hike through Icy Glen,[26] a small park preservation in Stockbridge. On another occasion, they attended a special fall pageant at the Stockbridge Town Hall. In addition, a rare total eclipse of the moon took place in November. An especially clear night, with little extraneous light, made the spectacle a most unforgettable sight.

---

[26] A nature walk through a glacial moraine

Work on the night before Thanksgiving ceased at 3:00 AM. Dan lived in Housatonic at Mrs. Smith's boardinghouse. When Dan invited Joseph to spend what was left of the night and have Thanksgiving dinner there, he eagerly accepted. To celebrate the holiday with turkey and all of the trimmings was a special treat because Joseph did love a good meal. At this dinner, Joseph first discovered real American apple pie.

Winter came early in December. It brought some very cold weather, including several days of temperatures dipping below zero degrees Fahrenheit. In a general store, Joseph was happy to find a pair of buffalo hide mittens with big cuffs, which kept his hands and wrists warm. They were one of the best investments he ever made. Though mended many times, they lasted most of his life. When the cold weather froze the ponds, he could skate a few times. Once or twice, he enjoyed tobogganing with his men.

By December 21, the trolley line from Great Barrington to South Lee was electrified. At Christmas, Joseph cut his first fir tree from the woods near Glendale. With the Chapins and a couple other friends, he celebrated the holiday, exchanging gifts and sleigh riding to view the holiday decorations. Joseph remained in the area until the end of the year. He did odd jobs while looking for work with Westinghouse Electric Company and others. However, without any further prospects, he returned to New York City. Another crew finished the trolley line south from Housatonic early the next year. People could finally ride all the way through Berkshire County, from Connecticut to Vermont. It operated successfully until 1930 when competition with the automobile made it obsolete.[27]

---

[27] O. R. Cummings, "Berkshire Street Railway," *Transportation Bulletin*, no. 79 (January–December 1972).

Chapter Four

The New American

Returning to New York City in 1903, Joseph didn't waste any time finding employment again. He was never one to scoff at any job, regardless of how menial, as long as he could pay his expenses. Thus, for a brief time, he worked for Mr. I. J. Flinn doing minor jobs, for example, making keys and small repairs.

Fortunately, he again lived with Fritz and Luise Hettrich. They had been looking for a new home for some time. In April, they finally moved from St. Ann's to Westchester Avenue. They were still in the Bronx. Joseph gladly helped them pack their household items and moved with them to the new address.

For several months, he had been corresponding with Mr. Wilson, his boss at the North Service Electric Company. Mr. Wilson had established his own electrical repair shop. Toward the end of January, he hired Joseph. Joseph initially worked on various tasks all over the city, including installing electric wiring at the Macy Square, Lexington, and several other hotels. He repaired electric machines at the Mott Street Iron Works and Brakridge Construction Company. He installed motors at the Bayer Factory, the Ladiesware Factory, and other similar jobs.

When Mr. Wilson expanded his business, Joseph made his debut into the technical side of the arts and culture by working in penny arcades. He repaired the fortune-teller, candy hoist, dancing lady, and other arcade machines in the city, New Rochelle, and Newark. As trusted employees, he and another colleague collected the money from the machines on Saturday night. He was later put in charge of counting the money and keeping the accounts.

On one occasion, when installing a machine in a new penny arcade, Joseph stepped on a nail that had not been removed from a packing case. It was stuck in the same right foot that had been injured twice before. His feet seemed to be the most vulnerable to accidents. When Joseph got home, Luise fixed a poultice for his foot. During the night, he soaked it regularly in hot water. For several days, he hobbled around in some pain until it finally healed.

After work and on weekends Joseph exercised. Skating was still high on the list of activities. Friends, such as Joe Klein and Joe's sisters, Millie and Jenny, often joined him. If time did not permit, walking was always an alternative.

When the weather turned warmer, Joseph was drawn to the water again. He took long walks to the North River or East River to watch the boats and ships as they plied back and forth within his view. By March, he decided he really wanted a boat of his own. He searched the boathouses and builders along the Harlem River to find one who could build him a boat within his budget. On March 14, he ordered a small motor launch from the Kingsbridge Shipbuilders and paid $100 as a down payment. On May 2, it was finished. With the final payment of $150, his heart raced with excitement as he put it in the water. He named his boat Der Satz.[28] During the maiden voyage on the next day, Joseph happily explored the waterways from Spuyten Duyvil to Pelham Bay. From that day until mid-October, he spent most of his free time on the water. He frequently took along passengers, including friends, relatives, and business associates.

Along with his love of physical activities, Joseph's thirst for knowledge was insatiable. He voluntarily pursued many learning opportunities. That spring, he and Joe Klein attended a series of lectures on electricity and other scientific subjects at St. Ann's Avenue School.

For a time, the Klein house became his second home. Joe's father was always eager to hear what new improvements and modern innovations the boys were working on. Joseph was happy to have someone who understood and encouraged his efforts. Even after Joe moved to Newark in July in order to be closer to his new job, the Kleins still invited Joseph for meals.

Mr. Wilson moved his office to Philadelphia in mid-July. Therefore, when the Mark Wagner Company offered better pay, Joseph switched jobs. Mr. Wagner also had a string of penny arcades. Many of his establishments included peep show machines.[29] These new machines were particularly fascinating to Joseph because they added a new dimension to his knowledge of performing arts. They showed more advanced moving pictures that told short

---

[28] This term has several meanings, including a small fry (as in young fish), a treat, or entertainment. It is unclear which meaning Joseph chose for the name of his boat.

[29] In 1894, Thomas Edison invented the Kinetoscope to show moving pictures. He had his own establishments to exhibit these machines. Several other inventors, like W. K. L. Dixon, also developed machines that appeared soon after in penny arcades and small performance halls.

stories. His work schedule was split between morning and evening shifts. During the morning, he usually repaired penny machines or phonographs, made sketches for new items, and took trips to suppliers for parts and equipment. Joseph had exceptional mathematical skills, and was completely trustworthy. All of the evening machine collections were brought in to him for the accounting. On many nights, he stayed until after midnight. Every day, he worked at least part time, including Sundays and holidays.

On most afternoons, he was free to enjoy boating in Der Satz, shopping, or sleeping. On rainy days when he didn't take the boat out, he'd go to the boathouse and talk with the owner, Otto Whit. The old boat tender always had an entertaining tale to tell of his early days as a seaman on the four-masted sailing ships, high seas, or foreign ports of call. Sometimes, it was just local boating gossip.

Despite the relaxation on the weekends, Joseph was having problems with his erratic schedule, which began affecting his concentration. He was becoming careless. One evening, while he was hurrying to get to work, he caught his foot between the hull and the bulkhead of the wharf when he docked his boat. The result was a sprained right ankle, the same one he'd injured before. This time, he bought a good pair of boots to provide support and protection to his poor ankles and feet.

Joseph was gradually given more responsibility. In the fall, his boss was sending him to city hall and the Board of Fire Underwriters for electrical permits. He ordered materials from the Douglas Company, Rosenfeld Manufacturing Company, and others. Jobs in the shop became more interesting as the items they worked on included electric signs, electric bells, sewing machines, electric communicators, and other newly electrified appliances.

One older man was apparently jealous of Joseph's ability to get along so well with his coworkers and the fact that management favored him with promotions and more responsibilities. The breaking point came when the boss, Mr. Mark Wagner, asked Joseph to work on his personal phonograph. The older man became really angry. He threw things around the shop, argued with everyone, and generally caused a lot of trouble. He was fired within a week. In contrast, a couple men on the evening shift, with whom Joseph worked, became quite friendly. Fred Kemper was one in particular. They began going out together for a meal after the accounts were finished at midnight.

Joseph turned twenty-one in July 1903. By then, he felt he had found his place in the New World. At the end of October, he had been in America for six years and was eligible to become a United States citizen. He hurried to file his application and picked up all the information he needed to learn to become a citizen. Because Joe Klein had moved to Newark, Joseph asked Mr. Weiss, one of his supervisors, to accompany him as a guarantor on the day of his appointment. Mr. Weiss agreed to go with him. On November 9, 1903, Joseph proudly took his oath of allegiance at the New York City District Court. As a thank-you present, he bought Mr. Weiss a table lamp for his home.

Joseph's trips to Newark suppliers became more frequent as the year ended, giving him more opportunities to see Joe Klein for lunch or a brief visit. Joseph also frequently visited his family on the weekends. In late November, he joined the skating pals, Martin, Henry, and Joe, for an outing at Van Cortland Park. That night, Joe got the flu and stayed home for two weeks. At the end of December, Joe Klein and several others from New York City were transferred to Philadelphia. Their boss, the same Mr. Wilson that Joseph had previously worked for, had established a new shop to service penny arcades.

Mr. Wilson invited Fred Kemper and Joseph to visit him in Philadelphia on December 30 and paid for all of their expenses. After lunching together, he showed the New Yorkers around the City of Brotherly Love. They saw some of the historic places, including Independence Hall, the Liberty Bell, and Betsy Ross's house. Hoping to entice the two men to work for him, Mr. Wilson also gave them a tour of his new shop, but he still could not match Mr. Wagner's salaries. Late that night, Joseph and Fred boarded the train and slept for a few hours in the Pullman car. Arriving in New York's Pennsylvania Station at 6:30 AM, they went straight to their jobs. That evening, Joseph celebrated his first New Year's Eve as a citizen of the United States by going out drinking with Kemper and Moses.

Adolph Zuckor and his partner, Mr. Goldstein, had just started in the penny arcade business with their first shop in Union Square. By the beginning of 1904, they were building new penny arcades and nickelodeons[30] in Philadelphia, the greater New York area, and Boston. Mr. Goldstein hired Joseph in February 1904 with a pay increase. The idea of a theatre that showed moving pictures intrigued Joseph. Though just recovering from a horrible cold, Joseph eagerly accompanied Mr. Goldstein to Boston on February 23.

---

[30] Nickelodeons were small motion picture theatres. The public paid five cents to see the show, hence the name. They presented short films, like the twelve-minute *Great Train Robbery*, *Solome*, and so forth.

They stylishly traveled in the saloon car on the Limited Express. That night, they stayed at the posh Crawford House. However, Joseph found lodgings at the less expensive Beacon Chambers the next day.

As part of the crew that installed new machines, he repaired and maintained them as well. The hours were flexible, even though he worked long hours and often stayed late into the night. During the middle of the day, Joseph had time to see his former landlords, the Bouckes, and other Boston friends. Joe Klein soon joined the crew. He was glad he was working with his old friend again. The job in Boston was finished in mid-March. Joseph was assigned back to New York City. Joe returned to Newark.

When he returned to New York City, the weather was getting warmer. Der Satz was once again his refuge. When he didn't have to go to Newark or leave town, he was at the boathouse during the afternoons that spring and summer. On one such occasion, everyone was talking about a horrible accident that had just happened.

"Disaster at Hell's Gate" read the headlines on June 19, 1904. It was the worst single disaster to hit New York City prior to September 11, 2001. A catastrophic fire broke out aboard the *General Slocum* steamship in the East River at the junction with the Long Island Sound. It shocked the whole city. Most of the 1,331 passengers were family groups of German descent. They were out for an annual Sunday school event. From the horrid accident, only about 300 survived. Fortunately, none aboard were Joseph's relatives. However, the event did hinder his own boating for a brief time because he often took his boat in the same area.

By the age of twenty, girls were becoming increasingly attractive to Joseph. He began thinking about a permanent relationship, but he wasn't ready to get married yet. He saw Jenny Klein, Joe's sister, several times. She was good company, and she enjoyed sports. They went many places together, including the American Museum of Natural History and the Metropolitan Museum of Art, which he would not have seen by himself. Jenny came along for boat rides to Clason's Point or up the Harlem, Bronx, or Hutchinson Rivers.

With his friends from work, like Chester and Andrews, entertainment depended on the weather and how much money they had to spend. Boating to explore Port Morris, North Beach, City Island, West Farms Creek, and other water-accessible places, sometimes included afternoon excursions to Dreamland and Steeplechase Park. In the evenings, they enjoyed shows at the Imperial Roof Garden, the Aerial Roof, and the Madison Square Roof Garden.

They occasionally took the ferry to see a belly dancer in Astoria. They once went to *The Wizard of Oz*, which was Joseph's first Broadway play. In addition, when the new subway line opened, Joseph had to immediately try it.

Joseph also loved reading. When Luise gave him a book of Henry Wadsworth Longfellow's work, he read it with great delight. It was a cherished addition to his growing personal library.

Joseph still loved music and wished he could be some kind of musician himself. He eventually purchased a banjo toward the end of 1904 and took a few lessons, but he never really learned to play. He always hoped he would have the time to master it, so he never got rid of it.

After receiving a five-dollar raise in pay, his income was good now. By the end of March, he had managed to save more than $500. He deposited the money in three different banks: the Bowery Savings Bank, Dollar Bank, and Boston Savings Bank. When the company's insurance agent came to the shop one evening, Joseph asked if he would be eligible for life insurance. The insurance agent assured him that he would certainly qualify and gave him the papers to read. A couple days later, Joseph signed the papers, making Luise his beneficiary. It was a $5,000 policy. In those days, that was a lot of money.

On many occasions, Joseph's work still took him to Newark. He managed to see Joe Klein for coffee or lunch. Occasionally, if he finished work early, he would visit his cousins, the Comisars, who would invite him for dinner. Joseph often saw his brother at the bakery, but Karl was always too busy to get away. Their visits generally were just brief conversations in the kitchen.

The machine collections in the late evenings plus traveling between various shops and new construction sites engulfed most of Joseph's time. He traveled to a new site in Baltimore or went to Boston to check on new shops and repair phonographs and various machines at other sites.

One day, while en route from Boston to Philadelphia, he didn't have time to go home. When the train reached New York City, he got off at 125th Street and went to a penny arcade where there had been a relatively small fire. No one was there because the insurance claims had not been settled yet. He slept for a couple hours on the floor before continuing his journey. Unwashed and wearing the same clothes, he arrived in Philadelphia to look at another site. He finally fell exhausted into his own bed in New York City at 2:00 AM.

Mr. Wagner was now working with Mr. Goldstein and Adolph Zuckor because they had bought out his company. New arcades opened on 125th

Street. One was on the East Side, which was renovated after the fire. Another was on the West Side. Two more were on Grand Street, and one was on 14th Street. Constantly consulting with three bosses kept Joseph running from one site to another. At this time, he began learning how big business was conducted. For instance, between 1900 and 1904 in New York City, when one bought concrete by the cubic yard, one was only given nine-tenths of a cubic yard. That was all they delivered. If one figured a job for ten cubic yards, one had to buy eleven or end up short. Nevertheless, that was standard operating procedure.

During the same time in New York City, it took quite a process to have plans approved. When his bosses sent Joseph to have plans approved for a new penny arcade, he was directed to a downtown city office. He left the plans with someone who mumbled he had come to the right place and would hear from them in due course. He then waited for the permit to begin the construction work. Two weeks passed, but nothing happened.

Joseph finally went to Mr. Zuckor and said, "I can't understand why those plans haven't been approved."

His boss laughed. "Go down there and take Mr. So-and-so out for dinner. Slip this fifty-dollar bill in his pocket. That should take care of him."

With Mr. Zuckor's money, Joseph did as he was told. Sure enough, the plans were approved the next day.

Work continued much in the same fashion into 1905. Joseph installed and repaired machines, such as the gypsy fortune-teller, electric rifle, phonographs, and film machines, for the penny arcades and nickelodeons in Manhattan and Newark. Then the Goldstein/Zuckor enterprise bought out Mr. Wilson.

At the end of February, Joseph, Chester, Philbrook, and Harry Weiss were transferred to 1221 Market Street in Philadelphia. Philadelphia offered opportunities to experience another part of American life. For the first time, Joseph tasted Chinese food because Chinatown was not far from where he was working. After settling in a boardinghouse on Race Street, he began going out with some of the new people who worked with him. They introduced him to minstrel shows, museums, and vaudeville at Keiths Theatre.

In Philadelphia, Joseph stayed in touch with his family in New York City and New Jersey via the telephone. In March, he learned of the death of Luise's mother, Mrs. Laier, thus ending the big large family gatherings. While the family still got together, the large parties were no longer held. The children were growing up, and the older ones were concerned with their own careers. Everyone had settled permanently in the adopted country and had made new friends. They no longer needed the social support of relatives.

A letter from Austria in February 1905 announced that Joseph's youngest brother, Edward (Eddy), would soon be coming to New York City. Unluckily, Joseph was already in Philadelphia before Eddy arrived at the end of March. Once again, Luise met him. On April 1, Joseph took off the weekend to see his sixteen-year-old brother. Eddy had grown into a handsome young man with a pleasant personality. He was staying with their brother Karl and his family. Over a couple German meals with Luise and the Fetzers, the family in America heard all the news from the family in Austria. Eddy was naive, but he was eager to learn how to survive in new surroundings. Joseph wanted to believe he'd soon adjust to life in America.

In Philadelphia, working on player pianos, storage batteries, key machines, and phonographs kept Joseph busy. With the warmer spring weather, leisure time was spent in Fairmont or Woodside Park or boating on the Delaware River. One Sunday, a group of his coworkers took a boat ride to Wilmington, Delaware. They had a nice seafood lunch and took the electric trolley home.

It had been more than seven years since Joseph had left Austria. He had been working steadily without any time off since then. Letters from home kept urging him to come back for a visit. Eddy had also delivered the same message from their parents.

Toward the end of April, Joseph talked to Mr. Goldstein and Mr. Zuckor about taking a vacation. He wanted to leave in two weeks. They agreed to let him go and said to check with them when he returned. On May 13, he left Philadelphia for New York City to begin arranging for his trip to Europe.

Chapter Five

The Journey Home

Over the next week, Joseph prepared himself to take the long journey to Austria. Joseph needed to get an American passport and buy a suitcase, presents for the family, and a ticket. He purchased a second-class passage on the Red Star Line's *Vaterland* for $266.80 and withdrew enough money from his accounts to cover the ticket and provide fifty dollars for incidental spending in Europe. The passport was ready on May 18, 1905.

Between chores for the trip, he managed to see his brothers, Uncle Fritz, the Hettrichs, Fetzers, and other relatives. All of them had letters and messages for families and friends in Europe. Karl also had a small photo of himself for his mother.

Joseph managed to visit people who had worked with him on 125th and 14th Streets. He spent one evening on Der Satz with Eddy, Andrews, and Gustof Walz. On another night, he sang with his pals. In the afternoons, he had coffee at Lindhof's with Luise and her father. He also had drinks or coffee with Eddy and friends.

On the day of departure, May 20, Joseph was up at 7:00 AM. He telephoned Eddy to say good-bye. At 10:30 AM, he, Luise, and her father went to the Red Star dock. Andrews and his sister were already waiting there. The four of them came on board with Joseph to look around the ship while he checked into his cabin, number 332. After the whistle signaled for all to go ashore, Luise and the others waited on the dock waving to Joseph, who was on deck. The stout tugboats effortlessly pushed the large ship out from the pier and guided her into the shipping lane of the North River. As they passed the Statue of Liberty, Joseph thought of the first time he had seen her. The feeling of warmth and anticipation he felt as a youngster returned. This time, he was sad for leaving her behind.

Once underway, Joseph noted some of the details about the ship. The *Vaterland* was one of the four twin-screw[31] steamers in the Red Star fleet. It carried 12,017 tons. Like her sister ships, she was 580 feet long and sixty feet wide. Her captain was R. C. Ehoff. These ships were the largest and fastest steamers between New York City and Antwerp.

The first call for second-class dinner was promptly at noon, even though they hadn't left the outer harbor yet. Joseph was assigned to table thirty-six. During the ten-day voyage, breakfast was between 8:00 and 9:00 AM. Dinner was at noon, and supper was at 6:00 PM. For those feeling hungry before bed, a night meal was served from 9:00 to 10:00 PM. Joseph, always ready to eat, was usually among the first ones in the dining room and enjoyed a snack in the evening.

Like his earlier voyage, he mostly spent the days on deck, watching for ships, icebergs, and so forth. This time, there seemed to be more sea traffic. He saw many large passenger steamships and freighters, but he didn't see as many sailing ships and smaller private yachts.

Looking for companionship, he read the list of his fellow travelers in second class. He might have noticed the name of the budding author, Joseph Conrad, but it was just another name to him. Several other young passengers were traveling alone, and Joseph did become acquainted with them, including Fred Koewing, Martha Rosenfeld, Rosa Straehl, and Jacob and Simon Shapinsky. With one or more of these companions, he walked the decks, attended evening concerts in the salon, or played various deck games. On Sunday, May 28, the amateur grand concert was held in the second-class dining room to aid the seaman's fund. Of his onboard friends, only Martha Rosenfeld and Joseph were brave enough to perform with the sixteen other artists. Martha recited the "Flower Girl," and Joseph ended the program with the "New England Farmer."

The captain congratulated all the performers, and the audience showed their appreciation with a rousing round of applause. After the show, Joseph and the other participants met in the bar for a congratulatory drink. For the first time, Joseph personally understood the attraction to the stage.

---

[31] Screw is nautical nomenclature for propeller.

The European coast appeared on the horizon on May 29. At 11:00 AM, the ship entered the canal leading to Antwerp. After the evening concert, Joseph watched the ship leave the canal and enter the harbor at 11:00 PM. On the morning of May 30, everyone bade farewell. At ten-thirty, they disembarked for their separate destinations.

Joseph went to the railroad station, bought a ticket, ate lunch, and boarded the train for Cologne. This was his first visit to Cologne. Because it is located on the Rhine River, he could not resist the temptation to take the boat ride down the river instead of continuing by rail. The famous cruise boats had evening schedules as well as during the day. There was just enough time to see the famous Dom[32] and a few other important buildings and museums in the city center before he boarded the boat. Once on board, he had something to eat and drink in the boat's cafe.

It was a beautiful, moonlit night. As the cruise began, he commandeered a deck chair to view the sights along the river. Relaxing in his chair, he saw Bonn, home of Beethoven, glittering on the west bank. He also saw the Bad Godesberg Castle ruins, Drachenfels,[33] and the ruined Königswinter Castle. They passed Nonnenwerth Island with its Benedictine nunnery and Remagen. Perched on the cliffs and crags on both sides of the river were Burg Ockenfels, Schloss Arenfels, and the Hammerstein Castle[34] ruins glowing eerily in the illumination of the full moon. The cruise then headed to Koblenz at the juncture of the Rhine and Mosel Rivers, where Eherenbreitstein Fortress stands guard from the eastern heights. After a brief stop to let people off and on for the journey to Mainz, the boat glided past Stolzenfels Castle, the Lahn River, Burg Lahneck, and Marksburg Castle,[35] which was the most completely preserved. The resort town of Boppard with its garlands of lights was spectacular at 3:30 AM. The river wound past Reinfels Castle ruins, St. Goarshausen, Burg Maus, and Burg Katz. As they safely cruised around the Lorelei rocks, Joseph smiled to himself, feeling fortunate the Rhine Maidens weren't serenading that night. Further upriver, the boat steamed past Oberwesel with its Schönburg Castle. At Mainz, opposite the juncture of the Main and the Rhine rivers, they docked to discharge more of the passengers. By five o'clock in the morning, they crossed the river to the final destination in Wiesbaden.

---

[32] The Dom is a cathedral that was started in 1248 and finished in 1880.

[33] From the Siegfried legend

[34] Home of Henry IV

[35] Now the home of the German Castle Association

The railroad station wasn't far from the dock. Joseph purchased a ticket to Bad Homburg, where his Tante[36] Katchia (short for Katherine) Faulhaber lived. She was a widow with three grown boys. Karl, the eldest, was living in New York City. Joseph was delivering a letter from him to his mother. He had to change trains in Frankfurt am Main because Bad Homburg is slightly northwest of Frankfurt.

Tante Katchia warmly received her nephew. She had not seen him since he was a little boy, but they had been in touch through letters and cards. Over a hasty lunch, Joseph told her about life in America. She was delighted to hear from her son and his family. Joseph filled in some of the details that the letter had not mentioned. He didn't see that family very often, so the information was mostly about New York City, the living conditions there, the food they ate, sports, and cultural attractions. He caught the 1:00 PM train to Frankfurt. At 4:26, he was on another train bound for Wien.[37]

In the café car, he tried sleeping, but he did not have much luck. When the train reached Würzberg, Joseph was just too stimulated to sleep. He was so excited at the prospect of seeing his parents, relatives, and friends again. When the train pulled into Wien, Joseph rushed through the station and hired a *fiacer*[38] to take him to Nüssdorfer Strasse, where his parents lived.

His mother had grown much heavier and looked older. At first glance, Joseph hardly recognized her. From the look on her face, Joseph must have also been somewhat of a surprise. He was no longer her fifteen-year-old boy. The hesitation was fleeting as mother and son embraced each other. She had not lost her laugh or affectionate manner. The first impressions vanished with hugs and kisses. The stream of questions about all of the family then came. His mother informed him that, except for Alma, the youngest sister who was away at a guild school for the year, all of the others were around.

Joseph gave his mother the small photos of Karl and himself that she had requested. She always wished to have all of her family home together just once. Unfortunately, she never got her wish. The closest she ever came was a photograph taken of those still in Europe in 1902. Karl and Joseph were already in America, so she wanted the small postage stamp photos to paste on the family photograph, as if they were pictures on the wall.

---

[36] Tante is the German word for "aunt."
[37] Vienna
[38] A horse-drawn cab in Vienna

Figure 12—The Franz Family. Pictures on the wall, Joseph and Karl.
Standing: Barbara, Ludwig, Maria, and Rosa.
Seated: Barbara Tinauer Franz, Edward, Karl Anton, and Alma

Karl Anton took time off from the office to greet Joseph with the familiar bear hug and pat on the head. He had hardly changed at all. He was a little heaver, but he was otherwise much the same as Joseph remembered. His comment on the mustache was, "It certainly makes you look older." Joseph couldn't tell if that was a criticism or a compliment. His father didn't stay long because he had to return to the factory. With a big smile, he promised there would be plenty of time for them to catch up with each other.

Later in the morning, his brother Ludwig's fiancée, Pepi (Josephine), came over. This was their first meeting, and she seemed like a very nice person. Joseph was glad Ludwig had found someone with whom he could share his life. They were to be married in early 1906.

Even though he was tired, Joseph felt compelled to see as much as possible right away, just to make sure things were as he remembered them. That afternoon, he briefly visited his sister Wetty and her husband, who was also Joseph's good friend, Karl Gumpinger. Unbelievably, they already had two little daughters. Katherine was two, and Aloysia was almost one. He left shortly after arriving, but he just wanted to let them know he would be around for a little while. Then he took a walk on the Ringstrasse to see some of the familiar sights, including the Rathaus, Burg Theater, his old school, the Staats Oper, and museums. He did all of this before returning home in time for dinner.

Over the next month, he spent many happy hours walking with his father and/or Ludwig. They talked about life in America, the family members living there, the family members living in Europe, the way the world was changing, the Panama Canal project, the completion of the Mount Wilson observatory, Theodore Roosevelt's accomplishment in mediating the end of the Russo-Japanese War, Ahmundsen's discovery of the North Pole, Sun Yat-sen's impact in China, and so forth. Joseph enjoyed hearing the European perspective and his father's prophetic warning, "Beware of the sleeping dragon," a reference to the Far East, particularly China with its tremendous population.

One day, his father took him to the Taussig Factory for a tour of the place where he worked. He was still the factory manager and proudly explained the process of making soap and scents as they walked from one room to the next. By the greetings received from the employees, everyone evidently liked Karl Anton. He also introduced his son to the owner, Mr. Gotlieb Taussig, who praised Karl Anton's knowledge and efficiency. He gave Joseph a book about the company, which had been printed in 1904 to celebrate fifty years of the company's operation. Karl Anton's picture was in it.

While home with his family, Joseph ate well. Vienna has always been famous for its pastries, *mit Schlag*,[39] and excellent cuisine. Joseph sampled all at every opportunity. He ate many excellent meals in fine restaurants, cafés, or homes of friends and family. When his mother cooked at home, he tried bringing something special from the market for the occasion, including asparagus, shellfish, pastries, and other things not part of their usual fare.

Ludwig and Joseph were probably the most alike in personality and interests. They enjoyed the same sports and loved music. They were interested in

---

[39] whipped cream

science and technology. They spent many evenings together, eating in small cafés, going to the Prater,[40] or walking to visit friends. On weekends, they went swimming or boating. Several times, Ludwig, Pepi, and Joseph visited their cousin, Anna Fitzing. The four of them went out together for coffee or a walk in one of the parks. Anna was good company, and Joseph always enjoyed seeing her. He visited with family friends several times. They sometimes had coffee. At other times, they met for a meal. People always wanted to hear about life on the other side of the Atlantic Ocean. Everyone seemed to have relatives or friends in the United States. They asked Joseph to deliver personal messages to one and all.

Karl Propst, a family friend, once came to the house with his young neighbor. Karl played the piano, and the young man sang for the family. It was a real treat to have such a lovely private concert. Afterward, coffee, wine, and homemade cakes and cookies were served.

Wetty and Karl were particularly inquisitive because they were seriously planning to immigrate to the United States in the near future. Karl would leave first to find employment. Wetty and the girls would follow as soon as possible. Joseph still felt closer to Wetty than any other family member and she felt the same toward him. She didn't have any qualms about taking her children to New York City because she knew her husband and favorite brother would be there. Joseph tried seeing one or both of them for a little while every day.

Joseph's sister, Marie, and her husband, Joseph Pruy, were living in Amstetten then. This was a long train ride, so Joseph was only able to see her a couple times. Marie always prepared a delicious lunch with Hungarian meatballs, a dish Joseph loved and had learned to make himself. She seemed happy enough, but she was concerned about her husband's prospects on the job. The Austrian economy was not very good, and they were thinking of possibly trying their luck somewhere else. In a letter sent later, she asked for Joseph's help. He wrote back, saying he would let her know if there were any possibilities in New York City. Things were discouraging because the United States was also in an economic depression then.

His sister Rosa was still unmarried. She had had a bad fall in the dress shop where she worked, but she had apparently recovered. Nothing ever seemed to go the way Rosa thought it should. When Joseph and his mother went with her to the doctor's office for a checkup, she complained she was broke because she

---

[40] Area of amusement park and sports fields

hadn't been able to work in a while. She asked her brother to give her some money to help her through this period. Joseph did give her twenty kroner and hoped that would be sufficient. He even took her to the opera with his mother. She seemed to think everyone in America was wealthy, so Joseph must certainly have lots of money, too. She never did understand that his trip was only possible because of long hours, very hard work, and careful saving. Moreover, it would take more hard work to replenish his savings.

The Danube River supplied many occasions of pleasure. One time, some of the family took an overnight boat trip to Krems and Stein. After a long climb up the Stein Mountain, they were rewarded with a gorgeous bird's-eye view of the great river and surrounding countryside. In Vienna, the Danube's banks were often the route of evening walks. On many afternoons, Joseph hired a small rowboat or shell and spent several hours on the river by himself. It wasn't hard to figure out that his love of the water had roots in this romantic river.

On July 3, 1905, Joseph paid for his return ticket on the *Deutschland*. He had reserved the ticket earlier and bought a railroad ticket for the trip to Hamburg. His father took him to Dandler's so he could buy a diamond. Stones were cheaper in Europe than the United States. Because Joseph was of marriageable age, his father suggested such an investment. He also loaned his son 167 Florins, so he would have enough money to get back to New York City. They said good-bye that night because Karl Anton had to be in the factory the following day. The big bear hug seemed longer this time, and the pat on the head was more like stroking. There was a definite sense of permanent separation.

The next morning, Joseph picked up his tickets, bought a German wallet, and had the last lunch alone with his mother before leaving. They went to the station together with Wetty, Karl, and a neighbor, Mrs. Konden. She tried, but his mother could not control her tears. As she kissed his cheeks, she held his face in her hands as if to permanently burn the image into her memory. Sadly, as time would prove, this would be their last farewell. The train left promptly with Joseph on board. The Fourth of July now had an especially special meaning for him.

In Prague, a change of trains took him on to Dresden and Leipzig, where there was a shorter wait between trains. He hadn't had any sleep until then, and he had very little to eat. When he got to the North Station in Leipzig, he hurriedly ate a bowl of hot, but rather tasteless, soup. In his haste to get on the train to Hamburg, Joseph caught his little finger on a small piece of metal that had come loose on the compartment door. He didn't think anything of it then, but it would become a problem over the next few days. Fortunately, he could sleep for most of the remaining rail journey.

The train pulled into Hamburg in the early afternoon of July 5. After taking his luggage to the baggage station for the shipping line, he checked into the Mosur Hotel, room thirty-nine, because boarding time was not until the next afternoon. He then looked around the place. They had a swimming room and nice Table de Hote dining room, where he finally ate a very good dinner. In the evening, he wrote a few cards and went to bed early.

In the morning, Joseph took a walk around the dock area where all the ferry trains came in. He also saw the Mulebur and Shaffer Hotels. After lunch, he left the Alt Stadt[41] for the harbor and boarded the ship. It weighed anchor at 4:30 PM before being slowly eased away from the dock and taken out into the Elbe River channel. Joseph stayed on deck, remembering his time in Wien and only half seeing the ship make its way down the river toward the North Sea.

When the supper bugle sounded at 6:00 PM, his little finger had begun to painfully swell. After supper, he got some iodine from the ship's pharmacy to put on the finger, but it began throbbing during the night. By 3:00 AM, sleep was impossible.

To try to obscure the pain Joseph read in the brochure about the ship. The *Deutschland* was one of the fleet of the Hamburg-America Line. Founded in 1847, it was the oldest navigation company in Germany. The fleet had 149 large ocean steamers and 174 steam launches, tugboats, and freighters. It was still building more. This ship was a twin-screw steamer, 686 1/2 feet long and 67 1/2 feet wide with 16,502 tonnage. Her fastest passage from Hamburg to New York City was five days, eleven hours, and fifty-four minutes. Mr. Kaempff was her captain.

After breakfast, Joseph went for a swim in the pool, hoping that would help the finger. At 10:00 AM, as the ship neared Dover, he saw the ship's doctor. The doctor told him that he had a bad infection and gave him something for soaking his hand.

On July 8, the pain in Joseph's hand was worse. There weren't anymore distractions, like taking on passengers in Dover and Cherbourg, to ease his agony.

---

[41] Old City

Sleeping most of the day did not help. This time, the doctor had to lance the pustule that had formed on Joseph's finger, which released much of the poison that had accumulated. The finger was still infected, but it felt much better. During the night, he still felt some pain. The dressing was changed the next morning. He could now at least take pleasure from food again and enjoyed the first concert on deck.

As on the voyage to Europe, Joseph enjoyed the scheduled meals, even though he could order additional food from 10:00 AM to 2:00 PM and from 5:00 PM to 10:00 PM. Drinks of all sorts could be had in the bar, which closed at midnight. Smoking was only allowed on deck or in the smoking room, which also closed at midnight. Elsewhere, lights were turned out in the public rooms at 11:00 PM. Cabins were equipped with private toilets, but bathing facilities were not provided. The latter were provided free of charge in convenient locations on each deck with help from an attendant.

After breakfast on July 10, the doctor lanced Joseph's finger again and dressed the wound with a salve and clean bandage. Joseph had to return that evening so the doctor could see if the medication was working. The swelling had decreased, and the hand was almost a normal color again. He was to continue using the salve for several more days.

Because of his finger, Joseph did not socialize as much as he had on the voyage to Europe. He only met one family, the A. S. Richters, including their son R. S. and daughter Clara. He also met one other man, F. L. Kramer. He undoubtedly did read the passenger list of both second and first class. The first-class roster carried the names of many well-known families, such as Steinway, Crane, Kennedy, Hammacher, Lynch, Merrill, Rothchild, and Scribner. The soon-to-be-famous film star, W. C. Fields, was on board with his wife. He would know two of the names, Stokes and Vanderbilt, more intimately later in his career. However, at the time, it is probable that he just barely recognized a few.

Fog obscured most of the view on July 11. Nevertheless, July 12 dawned bright and sunny. From Joseph's deck observation post, he spotted three steamships.

That evening, hundreds of lights and torches illuminated the afterdeck, making a festive atmosphere for dancing on the last evening. Joseph had never mastered the fine art of dancing, but he enjoyed the music and watching others as they whirled around the deck.

July 13 was Joseph's final day on the ship. They passed Sandy Hook and sailed into New York Harbor. They landed at four o'clock in the afternoon. Joseph had done a lot of thinking en route to America, including thoughts about his family, work, life, and future. If he had any lingering doubts about being an American, they were dispelled as soon as he saw the Statue of Liberty once again welcoming him warmly. This country was the land of opportunity. For better or worse, it was his home now, and he was here to stay.

Chapter Six

The Thousand-Dollar Table

Few of President Theodore Roosevelt's social reform efforts were yet felt that summer of 1905. In New York City, the effects of his earlier success at breaking the stranglehold of big business caused shrinking employment opportunities for ordinary laborers. For about a week after returning from Europe, Joseph looked for work in his field, but he didn't have any luck. Jobs in the electrical world were at a standstill, and none of Joseph's former contacts had any openings.

Not one to sit home and mope, Joseph saw friends and relatives, particularly his brother Eddy, and took people out for rides in his boat. His friend, Harry Weiss, was also unemployed. They often got together to talk about their mutual problem and possibilities for employment. Harry had heard the wholesale food business was prosperous. After all, people still had to eat. After some investigation, Joseph found that the S. N. Church Corporation, which was owned by Morris D. Kopple, was willing to take on partners in the wholesale grocery business. Five hundred dollars apiece would give each one a one-third interest in the business. Joseph had $200 in his bank accounts and could borrow an additional $200 from Will Klein, Joe's father, and $100 from Luise. Harry could not come up with any cash, so they drew up an agreement stating he would not draw out any money from the business until $500 from the profits would be accrued in his name.

The office was a small, dingy room at 136 Beekman Street. It had a couple tables. One was used as a desk and had a chair with it. The other was used for items that could be laid out or stacked on it. There were shelves for storage, a telephone, and one electric light. They needed something to transport the groceries, so they bought a secondhand wagon for fifteen dollars. With this office and equipment, they tried to be wholesale grocers.

Mr. Kopple was the president. Harry was the vice president, and Joseph was the secretary. Mr. Kopple was also treasurer and responsible for ordering the wholesale items. Harry was the salesman. After Joseph sent letters, he personally visited the individual grocers. Joseph kept track of the inventory and finances as well as all correspondence. He wrote form letters on the stationary they had printed to a list of grocery stores. Some of the commodities they sold—or tried to sell—were rice, tea, coffee, macaroni, and eggs. The first few weeks were frustrating because they didn't have many sales. Joseph was at the office early in the morning, usually by seven-thirty, to write the letters to potential customers and check the inventory. He gave Mr. Kopple the financial numbers from sales to enter into the books. New stock was usually delivered in the afternoon. Harry contacted new customers in the morning. He didn't come into the office until noon. By that time, responses from customers who wanted to purchase the products had arrived. Harry and Joseph would then deliver the goods. They ate a lot of macaroni and eggs to get rid of their first surplus stock.

In the midst of struggling to make the business work, his younger brother was unable to find work and kept expecting Joseph to help him. Joseph tried to look after Eddy and advise him, but he was tired of his constant complaints. Joseph finally told Eddy to do something for himself. On September 8, Eddy finally did find a custodial job with Jung's on Eagle Avenue in the Bronx.

By the second week in August, it was decided that the wholesalers were handling some of the wrong produce. Therefore, they switched to selling just teas and coffees and had a little better success. Joseph and Harry still weren't able to pay their debts because all of the income was used to pay for inventory and office expenses. No one was getting a salary. Because the office was in the Fulton Fish Market area, their lack of success was clearly due to the location. Harry was convinced they should be selling fish. After all, when Kopple bought the business, it had been the Eastern Fish Company.

In September, they again went to Mr. Kopple and told him they wanted to buy the S. N. Church Corporation outright and dissolve the partnership with him. They also wanted to lease his shop at 136 Beekman Street, the one in which fish were originally sold. Because Mr. Kopple's health wasn't very good, he agreed to sell his share. New contracts were drawn and duly signed on September 12. The first condition was that Harry and Joseph would be sole partners. Second, the purchase price for the fish company was $1,000, which was to be paid in installments of $50 a month. Joseph borrowed $300 from his friend, George Ernst Dahlhaus, and added his own twenty-five dollars as a down payment and one month's rent for the shop.

The fish business seemed easier. They only had to sell a load of fish for a fisherman and take five percent of the sale for themselves. The first contract was for a small amount of sea bass and eel, which sold very quickly.

Neither Joseph nor Harry knew a thing about marketing fish or seafood. Just because a few items seemed cheap, they made some bad choices. Clams were sold by the barrel, and they were a particularly bad risk. They spoiled very quickly. When Joseph got very sick on some bad clams, they learned to buy ice to keep the fish and shellfish fresh They did better with lobsters and various fish, including sardines, herring, and smelts. To get the best seafood, they learned to be in the Fulton market by four o'clock in the morning.

Harry Weiss had a good sense of humor. His gift of gab was a great help as a salesman. He was fun to have along on the boat or for an evening's entertainment or an afternoon at Minskys.[42] On the other hand, he didn't have any understanding of monetary affairs, so Joseph handled the bookkeeping. Near the end of September, Joseph realized Harry was not really putting in enough time to find markets for the seafood. He was rarely able to make the early morning hours at the fish market. He did little to change his habits, even after numerous discussions of the problem with him and his parents.

Harry's father, Dr. Ludwig Weiss, must have felt a little responsible because he visited Joseph about doing some electrical work for him. It was a welcome relief from the worries of the fish business and eventually paid some of his debts.

George Ernst Dahlhaus mercifully offered to take Harry's place in the business. Harry actually seemed relieved to get out of it all. He agreed to sell out for $100, which was about as much as he had earned toward his original contract of $500. On October 2, a new agreement was signed. Joseph had to borrow another sixty dollars from Joe Klein and took forty dollars from his bank account to pay Harry. In the next ten days, Joseph worked continuously to pay more of his personal debts. Fortunately, several loads of eastern herring were especially profitable.

Ernst, as he was called then, and Joseph became partners on October 12, 1905. He agreed to call his $300 loan the payment for his share of the business. Working with Ernst was much less stressful than with Harry, and they made some reasonable sales. The cooler season brought other varieties of fish to sell, such as mullet, cod, big head bass, freshwater fish, smelts, pollack, and steamer clams.

---

[42] Famous vaudeville and, later, strip show theatre on 14th Street and Irving Place

Through his contacts in the fish business, Joseph could obtain a dock pass. When Karl Gumpinger arrived on October 20, Joseph got a pass at the custom house to get onto the dock to meet the SS *America* when it pulled into Hoboken. He was happy to welcome Karl. He saw him through customs and booked him into a small hotel. The next morning, they had breakfast together. After work, they went to Luise's for dinner. On Saturday afternoon, they found a boardinghouse on St. Ann's Avenue in the Bronx. It was the same place where Joseph had stayed in 1901 and was much cheaper than a hotel. Karl moved in on Sunday. In the next few months, Joseph explained how to get around the city and got him started in English school. When Wetty and the children arrived, he would have to be well-established, so Joseph was happy to help. Besides, they really enjoyed each other's company. Karl later found a small apartment on the Upper West Side of the Bronx.

On November 7, 1905, Joseph could cast his first vote as an American citizen. This was a mayoral election, and he voted for George B. McClellan, the incumbent.[43] At that time, mayors served only two years. A popular mayor, McClellan was elected for three terms.

Christmas came and went without much celebration. No one had any money that year. Joseph worked right through the holiday and managed to see only a couple family members in the evenings.

Then, one shipment of cod was a disaster. They paid too much money for it. They had learned to put such excess in salt and ice in order to freeze and store it until later in the winter when the boats could not go out due to the ice. Fish were often scarce during these times. That winter was mild, and cod flooded the market. They knew they could not sell frozen fish at a higher price. Instead of paying to dump it, they practically gave away the lot. In addition, they still had to pay for their storage space in the 138th Street Market. With that, they lost what little profit they had begun to amass. The fish business was discouraging and unpredictable, and they didn't have any control over weather or market fluctuations.

Additionally, Joseph had a nasty sinus infection, which he seemed unable to get under control. By February, he just couldn't face the people who had loaned him money. In desperation, he went to his old boss, Mr. Goldstein, to beg for some kind of work, even part-time. To his delight, Mr. Goldstein hired him to start working in the automatic vaudeville shops as a repairman.

---

[43] From that time on, Joseph voted in every election, wherever he lived.

Ernst understood his partner's frustration, and Joseph assured him that he would still help in the fish business by doing the bookkeeping. Joseph's troubles didn't end there. Not a week after he started working regularly again, he woke up one morning to discover he had been robbed of everything in his cash box, including his watch, the diamond from Vienna, a bond, cash, and two bankbooks. Eddy's winter coat, his coat and vest, and a telephone set were also missing. Two detectives questioned Joseph about the loss and any people who had access to his belongings. They questioned everyone, including Karl, Luise, Fritz, and Eddy. They were suspicious of Eddy and investigated his activities for some time, but they never proved anything conclusive. The whole situation severely shook Eddy, and he left for Boston. Joseph had to go to the prison with the police to see if he could recognize anyone, but he didn't have any luck. He eventually got the bankbooks back, but that was all.

Adding to Joseph's misgivings about his future were two major stories that filled the newspapers in 1906. One was the incredible San Francisco earthquake and the firestorm it precipitated that destroyed most of the city. It was the single most devastating natural disaster the United States had ever experienced. Between 700 and 1,000 people lost their lives. Thousands more were injured. Property damage mounted into the billions of dollars. The other story that ran for weeks was about the notorious Typhoid Mary. A carrier of the dreaded disease, she had infected hundreds of people before she was finally found and arrested. The press played up the threat she posed with such melodramatic verbiage that many people in New York City wore protective face masks when leaving their homes.

Obviously, New York City was not the place to be. Joseph contacted Josh Lyden in Boston, a former employer. When the telegram came back from Lyden saying he was hired, he borrowed another twenty dollars from Luise. On March 18, he left on the midnight train for Boston.

As for the fish business, Ernst managed to keep it going for quite a while. Years later, he brought Joseph one of the tables from the shop. Joseph's only dividend from their $1,000 investment became his kitchen table.

CHAPTER SEVEN

Life-changing Experiences

Joseph was happy to be in the Boston area. He started working with electricity again in Woburn. He stayed in a boardinghouse that was near the job on Prospect Street in Woburn. His confidence returned, and life began improving almost immediately. Joseph's first experience with high-voltage power involved connecting 2,200-volt lines. He quickly learned that a shock from that much power could kill a person. One day, a short circuit created a terrible fire. Luckily, he jumped clear, escaping with only minor bruises and a scraped shin.

Though the job involved many long hours, Joseph found time to stop in to say hello to all of his old bosses, just to keep possible avenues of employment open. He also managed to see Mr. and Mrs. Bouke several times. They welcomed Joseph back with several dinners and evening visits. Joseph included Eddy on a few of these occasions. Eddy had finally found a job in the wholesale grocery market and was living in a room in South Boston. Though Eddy wasn't making a lot of money, he was at least supporting himself, which was a big relief to Joseph.

In Woburn, one of his coworkers introduced Joseph to a nice young lady, Ethel Gibson. She was a good-looking woman with a pleasant disposition. He enjoyed her company and took her out to dinner a couple times. The relationship was short-lived because work in Woburn lasted only a week-and-a-half.

He reluctantly moved in with Eddy for a couple of days until he found another job. This time, it was with Fred T. Ley, one of his employers on the Berkshire trolley line. Mr. Ley had moved to Springfield, but he had a contract with the Berkshire Electric Company. Once more, Joseph was on the move. He first checked in at the Berkshire Electric Company in Pittsfield. From there, he was sent to Lenox to take charge of the maintenance of power lines in that town. He was under the management of Benjamin H. Rogers. In Lenox, he first found lodgings and then introduced himself to Mr. Rogers. Shortly thereafter, Mr. Rogers bought the Lenox division, which became the Rogers Electric Company. Joseph was transferred to the new company.

Lenox was home to the majority of the wealthy Berkshire summer residents. Their magnificent cottages were larger than some hotels, and enormous, beautiful parks surrounded them. A recession may have been ongoing in other parts of the world, but there wasn't a lack of money here. These mansions and properties had been supplied with electricity since 1895.

William R. Stanley[44] and George Westinghouse settled their differences on the best type of electricity, AC versus DC and collaborated to put in the first AC underground system in America. It was installed by George Westinghouse, William R. Stanley and the Standard Underground Cable Company. Mr. Westinghouse was a major stockholder of the Standard Underground Cable Company. Rogers Electric Company eventually bought out the cable company, and Mr. Westinghouse became one of its directors. All of these people were among the founding fathers of the electrical industry.

On his "Erkskine Park"[45] estate, Mr. Westinghouse had a little workshop and powerhouse, that faced Laurel Lake. They were quite a distance from the house. With his money, he was able to hire William Stanley. and together, they installed one of the most sophisticated private lighting systems for his estate. His power plant generated the first alternating current. He had reflector lights outdoors that lit the tennis court and drive ways. Electric pumps supplied water to the artificial streams and fountains. Inside the house, he had installed the first indirect lighting.[46] The entire place was truly an electrical marvel. George Westinghouse's private steam and gas-powered producer plant was the first to supply 2000volt, 60 cycle power. It furnished the electricity to Lenox.

Joseph thought Mr. Westinghouse was an interesting man because he would recount stories of his years in the Civil War and how, after the war, he made his money with his invention of the railroad air brakes. Though they often differed on solutions to electrical problems or the possibilities as the electrical industry developed, he readily expressed his opinions on any of Joseph's problems. In some cases, Mr. Westinghouse proved to be wrong. He eventually lost the electrical battle to other companies that bought out his investments.

---

[44] William R. Stanley was the inventor of the high tension line carrying 2200volts, 2-phase electricity. The first line ran from Monument Mills to Great Barrington in 1893.

[45] The original house was razed to build one now standing at Foxhollow Time Share Condominiums.

[46] Approximately 1,800 bulbs were used.

One of Joseph's first tasks in Lenox was to locate cable troubles, which were as frequent then as they are today. In those days, the only way to locate a fault[47] was a complicated process. No one was ever sure how to find one. Additionally, the Lenox soil contained iron ore.[48] The duct used to carry the cable was made of iron with a cement lining. Under these conditions, detection was impossible.

After numerous failures in locating the faults, it became just a matter of guesswork to find them. A man doing this work would sometimes hold something, an iron bar or horseshoe, as a kind of divining rod. He would solemnly walk over the cable line. All at once, he'd say, "I believe I have it. Cut the cable right here!"

Joseph was amazed by the magic used to determine the spot. He immediately knew that one could not use appeals to the supernatural to locate the cable faults. He began working on a more reliable solution. The result was one of his first inventions, the Induction Explorer, a bundle of iron wires shaped like a horseshoe with magnetic wire wound over it. Both ends were connected to a telephone receiver. A few dry cells and interrupter[49] furnished the current that was sent over the fault, either between wires or between the wires and the cable sheath. This method greatly simplified the location of cable troubles. It was not until 1915 that this was mentioned in the technical press, and commercial "Fault Finders" soon came on the market.

Some of the business establishments in town were upgrading their illumination and equipment. Joseph's job included estimating the materials needed and assisting in the installation of wiring, fixtures, and switches. The Lenox Club on Yokum Road was an exclusive men's club where the wealthy gentlemen congregated to smoke their cigars and socialize. Lunches were occasionally served. Tennis and archery matches were held there, but gambling games were not allowed. Women were not allowed to join until much later. When they were, dinner was always served on Thursdays. This was the day when their cooks had a day off. Joseph was once again called to install adequate wiring. When the kitchen was expanded, he sold them the new GE stove, refrigerator, and other electric equipment.

---

[47] A possible break in the line

[48] Abandoned mine shafts from the former iron industry could still be seen in the area in 1908.

[49] A switch

Many large parties were hosted in the cottages.[50] Overflow weekend guests were often housed in the Curtis Hotel on Main Street. These accommodations were hardly as elegant as those in the grand houses, which could generally hold only a select few guests. Better lighting would help the competitive patrons prepare themselves more elegantly to meet their hosts. It would also help their daughters prepare to meet eligible men. Joseph supervised the installation of the new electric lighting in the halls and rooms.

Trinity Church on the corner of Walker and Kemble Streets was the place of worship for a majority of the wealthy. Their money had built the church, and many of the men served as vestrymen. The congregation donated money to electrify the church. This job was also done under Joseph's management.

Another job was to survey all the lines, make a map of the system, and install meters. Joseph had to gain access to all the homes that had electricity, mainly the big estates. However, this access was never through the front door. He could only enter through the servants' entrance at the back of the house. With 100 rooms, the largest place was Shadowbrook.[51] It was a timbered house on a field stone base built by Anson Phelps Stokes.[52] The family had turned it into an inn for a couple years when it became too cumbersome to maintain as a residence. In 1906, the place was sold to Spencer Shotter for a private estate again. Joseph repaired or replaced electrical fixtures and supplied additional electrical wiring in the house, barns, and outbuildings, as required for the new owner. Some of the houses, like Morris K. Jessup's Belvoir Terrace[53] on Cliftwood Street and the Schemmerhorn Cottage[54] on Walker Street, had their own storage batteries for auxiliary power. Joseph periodically needed to refill the batteries.

Whether installing the meters or later reading them monthly, Joseph made the rounds to Robert W. Patterson's Blantyre[55] and G. H. Morgan's Ventfort[56].

---

[50] For more details on the cottages, refer to *Berkshire Cottages: A Vanishing Era* by Carol Owen, published by Cottage Press in 1984.

[51] The house has been destroyed by fire.

[52] When Joseph returned to New York City, Anson Phelps Stokes' brother, E. D. Stokes, and his wife were on the *Deutschland*.

[53] The home is currently a summer arts camp.

[54] Schemmerhorn Cottage was razed for housing development.

[55] The home is now a very exclusive inn.

[56] It is now the present home of the Lenox Historical Society. It was used as a set for the orphanage in the film *The CiderHouse Rules*.

He went to the cottages of William Douglas Sloan's Elm Court[57] that seemed to be incongruously built and Samuel Frothingham's red sandstone Overleigh.[58] His monthly route also included the cottages of Spring Lawn[59] owned by J. E. Alexand and one of the most beautiful in Joseph's estimation, Giraud Foster's Bellfontaine.[60] where marble columns and statues adorned the roof and grounds. He was also at John Sloan's Windhurst,[61] the Grenville L. Winthrop Groton Place[62] and the W. B. O. Field Highlawn Farm[63] where he would make many other calls over the years. Joseph remembered these few the best because he'd later sell them the latest GE motors, kitchen equipment, and appliances. Many others in which he worked have been razed or lost to fire.

Even for the working class, life in Lenox was generally very pleasant. The proximity to the city of Pittsfield gave Joseph the opportunity to see many professional musical performances. On one occasion, John Philip Sousa and his band came to the Colonial Theatre. The convenience of the trolley line made it easy for Joseph and Ed Corbert, his friend from work, to get to Pittsfield in time to hear one of the concerts. Thanks to the patronage and/or participation of summer residents, various types of theatrical entertainment were presented at the Lenox Town Hall.[64] The first play Joseph saw was *Mrs. Wiggs and the Cabbage Patch*. Though this was an amateur production, it was done very well.

The Berkshires also offered Joseph opportunities to learn more in his chosen profession. Electricians in Pittsfield had formed a union and were urging everyone in the field to join. Even though he had been a member of the union in New York City, Joseph took the examination and was accepted into the local union on May 11, 1906. Many evening meetings and lectures were held in Pittsfield on the latest electrical regulations and inventions. Others covered the responsibilities and rights of the individual electrical workers.

---

[57] It is presently being restored for the second time as a bed-and-breakfast by Sloan's great-grandchildren.

[58] It is now part of the Avalon School.

[59] It is currently part of the Shakespeare and Company complex.

[60] It is now the exclusive spa, Canyon Ranch.

[61] It is now Cranwell Resort.

[62] It is now the Boston University Summer Music School.

[63] It is still an active dairy farm.

[64] It is now the Lenox Public Library.

The cottages employed many people. Many of these were young women who worked in the kitchens or as maids. It was a wonderful area to be a bachelor. Joseph didn't waste any time getting acquainted. One of the first girls he met was one of the maids, Gertrude Newberry. Gertrude lived in town, so she was usually free in the evenings. The tree-lined, unpaved streets winding up and down the hills in town were pleasant destinations for evening walks. Gertrude and Joseph took long walks. Sometimes, they just sat and talked on one or the other's veranda. Once or twice, Joseph even brought her candy. He occasionally talked with several other girls, but he never knew them very well and could not remember their names.

When Joseph started working at Springlawn, the Alexand house, on May 3 he met one of the maids, a lovely Swedish girl named Natalia Wahlstrum. He believed she was the prettiest girl he had ever seen. She had a twinkle in her eye that piqued his curiosity. Her soft accent stimulated what he had always felt was his basically unromantic nature. Sex he understood very well, but he did not grasp the meaning of a physical attraction that might lead to a long-term relationship. Natalia had a good sense of humor and laughed unaffectedly at the joy of just being alive. He fell in love with her the moment they met. He pursued her all through June.

Every evening, they took long walks and talked. She laughed at his stupid jokes. They were once caught in a short-lived thunderstorm. In the rain, they huddled together under a tree. Luckily, the tree was not the tallest one around, so they just got a good soaking as they watched the display of nature's fireworks. On Sundays, they spent more time together and ventured farther afield. They took hikes through Ice Glen in Stockbridge, had buggy rides through nearby towns, and went to the circus when it came to Pittsfield. On the Fourth of July, everyone in town was at W. D. Sloan's Elm Court to watch the marvelous fireworks show. As the bursts of colored lights momentarily lit the summer sky, Joseph was sure it was the best pyrotechnic display ever seen.

As summer progressed, so did Joseph's passion for Natalia. About the middle of July, they were exchanging hugs and kisses when they met and parted after evening walks. Their Sunday adventures together continued. There was once a band concert in front of the Curtis Hotel. Another time, Joseph hired a buggy for a ride to October Mountain. Then they went on a picnic and swimming in Lake Mackeenac. Joseph swam alone many times after work or between jobs, as the weather was very hot that July and August, but it was more fun having Natalia with him.

One evening, they attended a theatrical program in Sedgwick Hall. Joseph had helped to renovate the stage for that performance, and he ran the lights during the show. Natalia was sitting in the audience. She didn't say much afterward, but she apparently didn't enjoy being alone in the audience. This didn't seem to affect their relationship for very long.

August began with a personal catastrophe. Joseph was on his way to work in a wagon with the supplies needed for the Schemmerhorn Cottage. When they were almost at the gate, the horse tripped on a large rock in the road. As it stumbled, the load shifted. One of the front wheels ran into the same rock, and that was enough to make the wagon flip over into the ditch at the side of the road. Joseph was thrown out, along with all the electrical materials. The poor horse was so frightened that it bolted down the street, towing the broken wagon behind it. There was severe pain in Joseph's head, shoulder, chest, knee, and molar teeth. Fortunately, he didn't have any broken bones. One of the Schemmerhorn gardeners helped get the horse and unhitch the wagon. Joseph could ride the horse back to the stable. He was later compensated for his medical expenses, but he didn't take any time off from work. He desperately wanted to pay all of his debts in New York City.

Natalia was a great comfort in the evenings as the bruises and lacerations gradually healed. After a week, the aches and pains had subsided enough, so they could enjoy a couple more band concerts in Lenox. They hiked again through Ice Glen and took a rowboat ride on Lake Pontoosic in Pittsfield.

Their first serious disagreement came in mid-August. Like most lovers' spats, it was really silly. Joseph was handling the lights again for another dramatic production in Sedgwick Hall. Natalia didn't want to see the show. Granted, it was hot that week, and it was even hotter inside the hall. Joseph's feelings were hurt because Natalia didn't want to share something in which he was involved. Because he had committed himself to do the lighting for the theatre that summer, he did not think it would be right to quit. It continued to be a source of contention. Natalia was very jealous of his time away from her, but he had given his word to help. As far as he was concerned, it was a contract he simply could not break.

Tiffs were generally short-lived and soon forgotten. Within a day, they were walking in the woods or on Haven's or Lanier's roads. Sunday forays took them swimming to lakes in Pittsfield, to North Adams, and Berkshire Park. Natalia was friends with the Jenkins family. Mrs. Jenkins was also Swedish. Her husband, Allen, was caretaker of Bellfontaine. They had one son, Edward. The caretaker's house was a short walk down Kemble Street south of Spring Lawn and across from the Winter Palace.[65] Joseph and Natalia began visiting them occasionally.

On September 15, Joseph rented a horse and buggy and picked up Natalia for their first romantic weekend together at the historic Curtisville Inn.[66] The weather was still warm, so the drive was pleasant. The autumn leaves were just beginning to show some color, but most were still a dusty green. The old Stagecoach Inn straddled a small river, which originally led to a waterwheel that powered the first pulp paper mill in the United States. Lovely flower boxes on the balconies were blazing with the colors of the fall asters and marigolds. After dinner, Mother Nature happily supplied a crisp, clear sky bejeweled with millions of brilliant stars that rivaled the stars in the eyes of the young lovers.

Their relationship grew stronger in the next two months. They spent a couple more weekends together. Once, they took a trip to Albany. Another time, they went to Adams. Mostly, they walked and talked. They also visited with the Jenkins and others in Lenox.

In early October, George Dahlhaus[67] got married. With a check from his New York Life Insurance policy, Joseph sent George $100 of the debt he still owed him. In addition, he included ten more dollars as a wedding present. The fish business had finally begun to pay for itself. The next week, George sent Joseph eighty-one dollars from the profits.

The first week in November, most of the cottages closed for the winter. Laid off on November 8, Natalia had to return to New York City. They talked about keeping in touch. Joseph told her to let him know where she would be in the city. Joseph had his picture taken in Pittsfield just before she left so he could give her something as a remembrance. Then she was gone.

Joseph was suddenly alone again, and he didn't like it. Two days later, he had a letter from her. She said she had arrived all right and was staying with her friends, the Hoffmans. She missed Joseph very much and confessed her love for him. She desperately hoped they would be able to see each other again soon.

---

[65] It has now been razed.

[66] Today, it is Interlaken.

[67] After His wife liked the name George better than Ernst.

That did it! Joseph was never one to take long to react once he made up his mind. He knew he wanted to share the rest of his life with this woman. He sent her a telegraph informing her that he would be on the night train that arrived the next morning.

Early that Sunday morning, Joseph dashed from Grand Central Station to Luise's place to ask if he could bring Natalia for dinner, along with George Dalhaus and his wife. Luise agreed. He then met Natalia. He was so happy to see her. Immediately and without much ceremony, he asked her to marry him. If she accepted, he said he'd try to arrange for more time off or come down on a weekend in order to get married as soon as possible. With her delighted acceptance, they were officially engaged. They spent the rest of the day doing a little shopping as well as visiting Joseph's friends and relatives to introduce his fiancée. They formally announced their engagement during dinner at Luise's place. Joseph was glad to see George and meet his new wife. Joseph took Natalia back to the Hoffman's place and caught the last train to Lenox that night.

Two weeks later, he returned to New York City. Joseph bought a plain gold wedding ring, and he would've been happy with a quick civil marriage. He was a confirmed agnostic, partially due to things he'd seen en route to school as a youngster. He had to pass a Catholic convent early in the morning and would often witness a priest departing from the side door with a nun fondly saying good-bye. Sometimes, a priest was hastily adjusting his cassock. He knew this behavior was wrong and began questioning the teachings of the religion that permitted it. He had a strict code of personal ethics, but he was never associated with any organized religion once he left Europe.

However, Joseph knew Natalia enjoyed going to the Episcopal church and would want a religious marriage ceremony, so he called the Reverend Henry Marsh Warden at the Little Church Around the Corner to arrange a simple wedding. He then bought a bouquet of flowers and hired a handsome cab to take them to the church. At two o'clock on November 25, 1906, they were married. Afterward, they had a couple photos taken and sent copies to their parents.

Figure 13—Joseph and his bride, Natalia Wahlstrom

Joseph took the midnight train to Lenox alone because he hadn't found a place for them to live. They were not reunited in Lenox until December 4. Joseph had finally rented a small, furnished, two-room bungalow from his boss, Mr. Rogers. He bought a heating stove and wash basin. He then moved in his belongings. The ambiance was far from luxurious, but they were at least together and beginning their new life as a married couple.

Chapter Eight

The Saga of the Stockbridge Lighting Company

Stockbridge cottages were not as large as those in Lenox, and there weren't as many of them. Stockbridge was also proud of its colonial history and the local townsfolk had a lot more to say in the running of their town than the cottagers. Nevertheless, the summer people also demanded electricity, which proved to be fortuitous for Joseph.

Even though a financial panic caused a run on the banks in New York City and other corporate centers, the wealthy people in Stockbridge were not to be outdone by their Lenox neighbors. Inevitably, they demanded better electrical service. Charles S. Mellon, president of the New York, New Haven, and Hartford Railroad had purchased Council Grove[68] in Stockbridge. He was instrumental in getting the town electrified through his financial and moral support.

J. B. Hull owned the Stockbridge Coal and Grain Company. The Hull family actually ran several businesses in town. The old man never had much of an office. He hung out in Seymor's General Store[69] and took orders for coal in a pocket notebook. He drove down the street in a horse-drawn buggy, following behind the coal wagons to pick up any coal that fell off. He must have had such joy when he filled a paper bag. The Hulls were shrewd and made quite a lot of money. They got ten cents per ton for all the coal that came into Western Massachusetts, from Springfield to the New York state line.

Because of the family's wealth, Hull's son, Charles, was an eminent town resident. In late 1906, Mr. Mellon encouraged Charles to form an electric light company. After many meetings and discussions, Mr. Hull contacted the Rogers

---

[68] The house is on the corner of East Street and Main Street.
[69] At the present time, it is Williams Store on Main Street.

Electric Company to ask for their help. Rogers Electric decided to expand its area of operation and sent Joseph to explore the possibilities.

The town of Stockbridge insisted on control of its own system. The Stockbridge Lighting Company was formed with the proviso that the wires were to go underground. Though still employed by the Rogers Electric Company, part of Joseph's job was to design the electric lighting system for the Stockbridge Lighting Company.[70] Thus, Joseph had two bosses and double the work.

After the preliminary plans had been drawn up, it was determined that the center of Stockbridge would be wired underground. It would extend from East Street to Oak Lawn House, from Goodrich Street to Cherry Hill Road, from Glendale[71] to the Bowker's home and Daniel Chester French's studio. The cost would be $25,000.

Stock was offered for sale and $25,000 was raised with the guarantee of fifty customers. When the work started in 1907, there were actually fifty-two customers. Most of the year-round citizens of the town were either not aware of the project, not able to buy stock, or not interested in such a crazy scheme. The Stockbridge Lighting Company was a philanthropic venture of the more affluent summer residents. Some of those who bought the stock included Mr. Mellon (thirty shares) Mr. Choate[72] (ten shares), Mr. Woodward (thirty shares), Mr. Butler[73] (ten shares), and Mrs. Swan (ten shares). Mr. Hull was appointed president of the company and was given five shares. As for dividends, none was ever expected, but the Stockbridge Lighting Company did pay a dividend from the very first year of its existence.[74] Thus, the crazy scheme came into its own. Every year, electric distribution lines were extended, and more customers were added.

Power for the new system could be bought from the Monument Mills Company at four cents per kilowatt-hour. They owned a hydroelectric plant in Glendale and a second plant in the Furnace District.[75] A steam turbine plant producing 6000 kw was later added in the town of Housatonic. Based on the

---

[70] Mr. Westinghouse was also a stockholder in this company.
[71] One of the districts in the Stockbridge township
[72] A prominent lawyer from New York City
[73] Mr. Choate's law partner
[74] It is now part of the New England Electric System.
[75] It is another district in Stockbridge township where a blast furnace had once been located.

projected income from potential customers, Joseph decided the cost would be excessive if all of the wire went underground. Therefore, he planned two miles of overhead lines from the Glendale Powerhouse to Park and South Streets in Stockbridge. A cable would then go underground through the village. It would then go overhead to outlying districts.

There was great opposition to this plan because an earlier experiment by the Rogers Company had been a failure. They had run an overhead line from the Westinghouse Station near Laurel Lake, over Mr. Westinghouse's private property, about one-half mile to the main road.[76] It then went underground to Lenox village. In a lightning storm, the overhead line was hit, and the cable's insulation was punctured in many places. This incident convinced Mr. Stanley and Mr. Westinghouse of their mistake.

Obviously, Joseph was apprehensive about connecting overhead to underground. First, he consulted the National Underground Cable Company, whose cable he had priced.

They answered, "We cannot guarantee any cables if you connect them to overhead lines."

Naturally, Mr. Rogers only favored Westinghouse Company equipment. When asked about the reliance of their lightning arresters, Westinghouse replied, "We can give you a lightning arrester, which we believe to be the best thing available, but there's no guarantee it would give protection in your case."

Their low-equivalent arresters were the same as the old non-arcing spools that were connected in series with shunt resistance across the spools. Another consultation to protect the lines from electrical storms was with Stephen D. Field, the famous electrical engineer.[77]

He counseled, "Well! I wouldn't worry too much. If lightning gets in, all you have to do is fix the cables."

With this bit of encouragement, the arresters went in. Choke coils, which were then thought to be very beneficial, were also installed. To be doubly sure, two regular sets of arresters were put in multiples with the low-equivalent sets. Connecting the overhead lines with the underground cable was started just after Labor Day in 1908. Unhappily, there was a bad thunderstorm that

---

[76] Now Route 7

[77] Field was the director of Quartrulles Telegraph & Atlantic Telegraph Instruments.

evening. A lighting bolt broke the exposed cable near the Searing's house on South Street. Fortunately, the power had not been turned on yet, so it was just a matter of cleaning the joint and reconnecting the cable to the overhead line. However, after the power was restored, Joseph spent many long, anxious hours during lightning storms in the substation on Park Street. He watched the arcs across those arresters[78] until he could taste the nitrogen in his mouth.

Figure 14—The little electric substation on the corner of Park and South Streets in Stockbridge, Massachusetts

Happily, the system never failed!

As late as 1920, long after Joseph had proven its feasibility, George Westinghouse refused to have overhead lines connected to underground in Lenox. As a result, potentially profitable electric development in that town was

---

[78] See articles in *Electrical World*, published July 22, 1909, and December 2, 1909.

negated. For example, Lenoxdale asked for electric service from Lenox around 1914. Joseph designed an overhead line to run from a point on Depot Street in Lenox to Lenoxdale. There, it would connect to underground cable through the village.

When the proposal arose at the director's meeting, Mr. Westinghouse said, "Gentlemen, you have a good system. Keep it so!"

It did not pay to go underground all the way to Lenoxdale. The people there were not multimillionaires, like Morgan, Sloan, Lanier, Foster, and so forth. They could pay out of their own pockets for the service. Therefore, Lenoxdale went to the Pittsfield Electric Company for their power. Lenox lost the income.

In early 1907, the fatal accident of the electrical engineer, Frank Pope, convinced Joseph that grounding primary lines was imperative. A transformer breakdown in Great Barrington sent the primary voltage into Mr. Pope's house wiring. When he turned on an electric light in his cellar, he was instantly killed. This incident was greatly debated in articles of the technical press of the time. Joseph proved his theory when he installed the Stockbridge system. Not only did he ground the primary system, he grounded all of the secondary lines to buildings as well. There were years of apprehension and discovery as Joseph experimented with the potential power of electricity. The Board of Fire Underwriters still did not permit grounding. Insurance companies said that any fire resulting from such grounding would bring legal suits against the lighting company for damages. Nevertheless, the grounding in Stockbridge never caused a fire, and no one was ever hurt. It wasn't until around 1915 that the underwriter code made grounding permissible. Circa 1920, it was made mandatory.

Electricity is certainly not without danger, as Joseph found out the hard way. While working on the Mellon property in Stockbridge, specifically setting new lines on poles, the wires from Mr. Mellon's private, battery-powered transformer were being replaced with new lines from the town system. Joseph mistakenly picked up one of the old wires, which had dropped to the ground. The shock knocked him out. When he came to, he realized the power had never been turned off in the old system. Miraculously, it did not kill him, but stars spun around in his head for several seconds. One week later Robert Scott[79], an electrician from Westinghouse, and Joseph disconnected the old transformer.

---

[79] Robert Scott later invented the two- and three-phase transformers.

In the late fall of 1908, electric streetlights were installed in Stockbridge, despite many objections from the summer colony who wished to keep the village quaint. The power cables for the lights were installed in tile conduits with a drainage channel on the underside, allowing any water to flow into sumps along the route and prevent moisture damage to the line.

An unforgettable event occurred late in November when the streetlights first went on. Fred Aymar, chairman of the board of selectmen in Stockbridge, Charley Hull, and Joseph inspected the lights in Charley's secondhand car, a two-cylinder Maxwell with the crank on the side.

Charley Hull tried living stylishly in a big house on Main Street, but his father tightly controlled his finances. In those days, Charley seemingly had everything secondhand. Beside the auto, he had a secondhand house, a secondhand safe, a secondhand wife, and a secondhand baby.

The secondhand auto proved to be a problem on the day of the inspection. A light covering of snow was on the ground. The tires on the old car were absolutely smooth, causing the car to skid on a slick spot. It stalled. Mr. Aymar and Joseph started pushing. Mr. Aymar took a hold of the spokes, exactly like one would do with a horse-drawn wagon that was stuck in the snow. When Charley let out the clutch, Aymar went flying through the air. Fortunately, he suffered only minor bruises, but the inspection was over.

Another project was to lay out a street lighting system for the town of Lenox. By 1910, the Rogers Electric had become the Lenox Electric Light Company. The new manager was T. J. Newton, a typical New Englander. Because it was necessary to dig up all of the streets to install the streetlights, Joseph planned to put in a spare duct because there was only one. An extra duct would have been an advantage for replacing a cable and future secondary cables.

Mr. Newton asked, "What do you need an extra duct for?" Joseph answered that a new cable could be pulled in before the old one would be taken out.

He asked, "How long will the old ducts last?" Joseph answered that they could probably get along for ten or twenty years.

Mr. Newton replied, "That's long enough. Let the young squirts take care of that when they come along."

In his case, he did not last ten years. If all of the predecessors in electricity had such a philosophy, where would we be today? Joseph was not there to install the lights, but he had ordered all of the materials for the job.

CHAPTER NINE

A New Family and Social Life

The Rogers' house, where Joseph and his family were living at the time, was actually a small summer bungalow and was not insulated for cold weather. The winter of 1906–1907 was very cold. Temperatures were below zero degrees Fahrenheit on many nights. The little house was hard to heat. Natalia was pregnant and caught a very bad cold. In early March, after buying some new furniture, they moved into a small house owned by Mr. Würtzbach. They tried fixing up the house to be presentable enough for entertaining some of their new Lenox friends. However, Natalia was unhappy there because the place was still very primitive. In April, they found a nice place to rent for fifteen dollars a month on Church Street that the Whites owned. Once again, they packed their household effects and moved.

In the spring, the work in Stockbridge consumed most of Joseph's time, keeping him away from home longer. Decorating the Church Street place continued for some time before Natalia was satisfied. Being a very sociable woman, she easily made friends and was a capable hostess. She easily adapted to domestic life and enjoyed attending Trinity Church, where she met more people. On Sunday afternoons, she and Joseph often exchanged visits with Mrs. Grogen, the Ducots, or the Peters. Johnny and Mrs. Johnson, friends from New York City, spent ten weeks each year in a small bungalow near the lake. Joseph's boss, Mr. Rogers, and his wife also invited Joseph and Natalia for dinner once or twice a year.

Natalia regularly visited Dr. Hale as her pregnancy progressed. At eleven o'clock on May 27, 1907, a daughter, Natalie Caroline, was born at the House of Mercy[80] in Pittsfield. She weighed seven-and-a-half pounds and was nice and plump. Both mother and child were doing very well, but women were kept in the hospital longer after a birth then. Joseph could not bring them home until June 16.

---

[80] Now Pittsfield General Hospital

It is never easy being a new parent. Like most newlyweds, they were unprepared for the sleepless nights of the first weeks. Every cry incited a new terror until they discovered the reason for it. Because Joseph had so many younger siblings, he thought he knew what to do with babies. It is somehow different when you are the one with full responsibility for the infant's welfare. Joseph tried helping as much as possible, even though he often worked very late. He sometimes was away all night bringing electricity to Stockbridge safely and swiftly.

Mr. Hull, the Stockbridge boss, stressed, "The summer folks wanted their electric lights."

Little Natalie progressed quite normally during the summer. She eagerly took her mother's milk, gained weight, had her first outing on the Fourth of July, and slept through the night at three months. When her teeth started coming in, she was quite naturally cranky and hard to control.

Natalia, on the other hand, was not doing as well. On many days, she didn't feel well enough to face all the work that was necessary. One of her friends, Mrs. Johnson, was kind enough to help with the baby for two weeks in June, which gave Natalia some rest and adult company. She sometimes even stayed overnight. Natalia and Joseph were especially appreciative, considering Mrs. Johnson also cared for her own small son. Natalia's health still did not improve. She had several colds and abdominal pain. After a thorough examination, Dr. Hale finally diagnosed her problems as stemming from a fallen uterus. He could only prescribe rest and told her not to wear a tight corset, but this didn't help.

Small, as well as large, difficulties plagued Joseph and his family. In early June, Joseph was working at Elm Court, W. D. Sloan's house on Old Stockbridge Road, repairing an electric wall lamp. His ladder accidentally hit a nearby table, jarring a vase that stood on it. To Joseph's horror, the vase fell to the floor and smashed into pieces. He told the maid what he had done, but he didn't hear anything more about it until the end of the summer. When Mr. Rogers asked if he had truly broken the vase, Joseph recounted the incident. His pay was docked ten dollars to pay for the damage. It was an insignificant amount for the Sloans, but it was almost a half-week pay for Joseph.

To reduce some of his own frustration, Joseph began gardening. That summer, he started his first garden. It was mostly flowers and a few vegetables, but he enjoyed working in the dirt. It gave him something relaxing to do at home. Gardening became one of his regular summer activities, allowing him to get away from the stress of work and family life. Planting, weeding, or watering helped clear his mind of prejudicial jabs, lack of recognition for his abilities, and hurtful inferences he faced as a foreigner working in a closed society. The crops he grew included beans, tomatoes, and carrots. He also sometimes planted potatoes, peas, beets, turnips, and corn. Most importantly, they never talked back! The garden was also beneficial because its produce helped support the food budget. The surplus was canned or preserved and lasted into the winter.

For several years, he also raised chickens, which supplied fresh eggs and good chicken dinners. He once had a hen that sat on twelve eggs and hatched all except one. She was a great mother!

The Lenox house had a sprawling, old apple tree, which produced a large crop of apples each year. Unhappily, some of the apples fell over the neighbor's fence, causing a continuing dispute. A physical altercation erupted one year. In the scuffle, Joseph dislocated his left shoulder, further aggravating the situation.

As the weather grew hotter, Joseph went swimming and boating at Lake Mackeenac. He would sometimes go alone on his way home after work. Other times, he went with friends or colleagues. On Sundays, the family occasionally enjoyed picnics and swimming at one of the many lakes or ponds in the Berkshires. If time did not permit a trip to a lake, Joseph just sunbathed in the back yard.

Friends in New York City occasionally came to the Berkshires for vacations. In August 1907, George Dalhaus and his wife, Carrie, stayed with them for ten days. Joseph enjoyed seeing his old partner. In the evenings, he showed George around the area. They swam in the lake, picked wild raspberries, went to concerts, or just meandered on long walks. On one evening, Joseph took George to the electrical shop to see his plans[81] for the Stockbridge Lighting Company. The women and baby occasionally joined the men, but they mostly stayed home and talked or took short walks around town.

On August 9, a friend watched the baby so the adults could have an afternoon alone. Joseph hired a horse and carriage from Michael Prout to take George and his wife sightseeing. They were traveling south on Route 7, heading for the center of town. A bee apparently stung the horse because it suddenly bolted, throwing out Natalia and Joseph near Tanner's house. Down the road, George and his wife were thrown out on Christian Hill. None of them

---

[81] Joseph had drawn up all the plans for the Stockbridge Lighting Company's underground cable and overhead lines.

was badly hurt, but their clothes were soiled. Carrie, who landed in a puddle, was covered with mud. News of the accident even made the Eagle[82] the next day.

The last day of the visit, they lit a fire in the fireplace and toasted marshmallows. This was the first time any of them had tried this summer treat. Everyone had many laughs when the marshmallows got too black or fell off the sticks. All of them agreed it was an experience to be repeated.

October 20 was Natalia's birthday. Joseph cooked breakfast and brought it to her in bed. As a birthday present, he promised her a trip to New York City as soon as she felt better. On October 26, the family boarded the early morning train to New York City. They first went to John Johnson's apartment. Mrs. Johnson greeted them warmly. She showed off Elias, their one-year-old son, and served lunch. After lunch, they found a small room for reasonable rent and left their luggage there. They spent the rest of the day visiting Natalia's friends. It was midnight when they returned to their room. Little Natalie was remarkably good and slept much better than her father did. The next day was Sunday, so they were able to see more people, starting with the Johnsons. Johnny was home this time. They enjoyed another good lunch together with spirited conversation. They spent the afternoon with Luise and Fritz Hettrich. Rosy and Cary (Carol) Fetzer joined them for dinner. It was a delightful evening. Luise fixed a big meal with lots of good food for all.

On one night, they left the baby with the Johnsons and went to the Academy of Music to see *The Lion and the Mouse*. On Wednesday, Natalia and her friend, Ellen Hoffman, went to a matinee performance at the opera. On several evenings, the family visited George and Carrie Dalhaus. Joe Klein, the last of the bachelors, was included.

Natalia managed to go shopping and bought an overcoat and hat, among other things, while Joseph looked after the baby. They had a family photo taken with little Natalie on her father's lap.

The last day, while Carrie Dalhaus and Natalia went to buy a dress, Joseph took his daughter to visit the Fetzers and his sister, Wetty, and her two girls. Wetty had arrived in New York City during the spring of 1906. She and her family had moved to St. Ann's Street in the Bronx, not far from where Luise

---

[82] The *Berkshire Evening Eagle*, published in Pittsfield, Massachusetts

and Fritz had lived when Joseph first came to the city. This was his first chance to see them, so it was a very delightful reunion. That evening, Natalia bought cake, milk, and cookies to help entertain the Dalhaus family and Carrie's brother, Mr. Hoffman, and his wife, Ellen. Before boarding the train to Lenox on November 2, Joseph managed to buy a pair of shoes for himself. Thus, their vacation in New York City ended.

Two days later, Joseph proudly voted in his first presidential election in Lenox. His choice, William Howard Taft, won, becoming the twenty-seventh United States president. Little Natalie's first Christmas followed shortly after. They celebrated the holiday with Ase, a Norse goddess, on the top of the tree and gilded nuts as ornaments. Mr. and Mrs. Rogers sent presents for everyone, distributed very formally by their young son, Edward. The Franz family sent back homemade cookies and other sweets.

1908 was eventful in many ways. The first Model T Ford was produced. Wilbur Wright flew thirty miles in forty minutes. Grover Cleveland died. However, for the Franz family, it was much the same as the preceding year. Natalia was pregnant again. She had a very difficult time due to her previous uterus problem. She was in continual pain, and there were other side effects, such as nausea and chills. Additionally, she also had her hands full while trying to care for a young daughter. Natalie took her first unaided steps on August 10. From then on, she was off and running! Joseph often made breakfast and dinner, did the wash, and helped with the other housekeeping chores, even though all of his work was now in Stockbridge. His hours were very erratic. On many evenings, he worked overtime, which only compounded the stress on his wife. Sundays were the only days he could spend time with the family.

Families of friends were beginning to increase as well. George and Carrie Dalhaus had a little girl on February 26, 1908. In June, they came for another visit. Carrie and the baby stayed for two-and-a-half weeks, but George had to return to New York City after the first week. He came back for the Fourth of July weekend. Then all of them left. It was enjoyable having them around. The men got out in the mountains again to pick wild raspberries and blackberries. They could talk man-to-man without the women or children interrupting them.

John Johnson and his family rented a bungalow for August and part of September. During that time, they had lively visits with John Johnson and his wife. The Franz family helped celebrate Elias Johnson's second birthday on

September 6. It was fun to watch Elias and Natalie playing well together, as they were just eight months apart.

Joseph was very proud with the birth of his first son, Russell Karl Franz, who was born at home at four-fifteen in the afternoon on September 25, 1908. He weighed eight-and-a-half pounds. Miss Davis, a nurse, assisted Dr. Hale. Natalia stayed in bed for ten days, and Joseph hoped her physical problems were over. Regrettably, Russell couldn't take his mother's milk after the first month, causing trouble for both mother and child. There was also a problem with Natalia's teeth. She was suffering from a lack of calcium. It was common for mothers to have toothaches or even loose teeth. People once said "one tooth for every child," but no one knew the cause. Natalia did eventually loose a tooth and had another one filled.

After the baby started on a milk substitute, he developed a terrible diaper rash and a ruptured navel. His parents were both exhausted from lack of sleep and too much work. To relieve the stress, Joseph finally hired a girl to help Natalia for three months after Russell's birth.

Because the children and Natalia were particularly prone to colds, the house had to be heated at all times. When the weather turned colder, they had to light fires in the fireplaces and stoves to keep warm. If life was not difficult enough, Natalia was desperately trying to start a fire in a stove one cold November day. As a starter, she used a little kerosene. In the resulting flash of fire, she was badly burned and had to call a doctor for help.

The Christmas and New Year holidays passed relatively routinely. Joseph cut a tree from the woods, and it was trimmed on Christmas Eve. They exchanged a few presents. Natalie received a doll from the Stockbridge boss and his wife, Mr. and Mrs. Hull. Russell received a jacket from the Lenox boss and his wife, Mr. and Mrs. Rogers. Surprisingly, a package from Bratislava, Czechoslovakia, arrived. It was from Joseph's sister, Marie. She and her husband had moved there and established a small *Gummi Fabrik*.[83] She had sent some of her own fruit preserves. For the next couple years, they continued to enjoy her tasty treats that came for Christmas.

At the beginning of 2009, thanks to a sterilizer for his bottles that was purchased from Mr. Hull, Russell was doing better. By the end of the month, he was finally sleeping through the night. In the middle of February, he was hold-

---

[83] Gum or India rubber factory

ing his bottle by himself. Even though he seemed to have some kind of congestion or cough all winter, he, like his sister, grew bigger. He had cut his first teeth. By his first birthday, he was crawling so fast that he was becoming hard to catch. At fourteen months, he was walking.

Figure 15—Joseph, with Russell (fourteen months) and Natalie (two-and-a-half years)

The children were a constant source of pleasure as they developed new skills. Natalie loved the snow. Joseph regularly took her when he went to the post office or got the newspaper. In March, Joseph and Natalie went ice-skating together for the first time. Natalie was afraid of falling, so Joseph constructed a little support that she could hang onto and push as she walked on the ice with her double-runner ice skates.

When the spring chickens hatched, Joseph brought one of the chicks into the house for the children to see. Natalie wanted to keep it for herself. She was unhappy when she was told it had to go back to its own mother. Everyone had fun when they went to pick wild raspberries and blackberries in the woods. All of them ate as many as they picked. In the summer, Natalie had her first haircut and went swimming with an inner tube in Hager Pond.

When swimming at Hager Pond one time, Joseph ran into a tree branch, injuring his left shoulder. As a young man, Joseph's feet and ankles seemed to be the weakest parts of his body. His left shoulder was now his nemesis. In his twenties, he had dislocated his left shoulder. There was also the incident with the neighbor over the apple tree. In December 1909, when he was installing a switchboard in the Furnace District Powerhouse, Joseph somehow managed to dislocate his shoulder for the third time. It was extremely painful that time. The medical book that Joseph had purchased after Natalie was born recommended a hot sugar poultice. It helped relieve the immediate pain, but the shoulder was still sore for several days.

His wife seemed to be in better health. Mrs. Carlson came from New York City for a ten-day visit in August. She and Natalia had become friends when they both worked at the Alexand house. They had several mutual friends in Lenox, whom they enjoyed visiting together. Natalia enjoyed the visit. Her strength gradually returned until she developed anemia in September. The doctor prescribed a regimen of iron pills, but it took another couple months before she was back to normal. George Dalhaus came one Saturday while Mrs. Carlson was there, but he did not bring the family. He was hoping he and Joseph could have a couple days alone, away from sleepless nights and diapers. Finding only women and children on his arrival was disconcerting. When Joseph got home, George asked to be taken to Stockbridge in time to catch the early train on Sunday morning. Joseph sympathized with his friend. He knew all too well how hard it was to be a parent. He was sorry he could not help George more.

That fall, business was hectic, and several large sales were on the line. Mr. W. K. Vanderbilt had expressed interest in several electrical additions for his Berkshire estate. After several weeks of preparations, Elmer Newton,[84] the new assistant manager, and Joseph were to make the sales presentation. The first snow of the year fell on October 16, the day Elmer and Joseph were going to New York City. Elmer was being groomed to succeed his father as the manager of the Rogers Company. This sale was large enough to impact the company's bottom line, and Elmer was expected to close the sale that day. Unfortunately Elmer was too overwhelmed in the great man's presence and didn't utter a

---

[84] Elmer was the son of T. J. Newton, manager of the Lenox Lighting Company.

word, so Joseph did all of the talking. As a result, Mr. Vanderbilt ordered a new battery and wanted to see plans for the installation work on Monday.

    This decision, required Joseph and Elmer to stay in the city. Joseph took advantage of this to schedule a few visits with friends and family. Joseph left Elmer and went to the Dalhaus apartment. He was pleased to find George in a happier state. They invited him to stay with them, which he gladly accepted. Then he went to see Luise and Fritz Hettrich. From there he went to Elmer's hotel to put the finishing touches on the battery drawing because Elmer couldn't draw a straight line with a ruler. He made it to George and Carrie's apartment at midnight. On Sunday, he awoke early and went to Otto Whit's boathouse to say hello to old friends there. He had sold his boat to help pay some of his debts, but he still had fond memories of his time on the water. Next, he went to Second Avenue to buy an overcoat for four dollars. After that he visited his cousin Valeria, who was now married to a Mr. Ullman.[85] He had lunch with the Hettrich family and briefly visited with Wetty and the Fetzers. That evening, Joseph stopped at Mrs. Carlson's place before meeting George's oldest sister, Carrie and George for dinner at the Schnitzers. He didn't sleep too well that night because of the fussy baby. On Monday morning, Joseph met Elmer at the elevated station near his hotel, and they went for breakfast together. They walked around the city until noon and lunched in a restaurant near Grand Central Station. After delivering the plans to Mr. Vanderbilt's office, they caught the afternoon train, arriving in Lenox at eight o'clock that night.

    The early snow was gone by then, but the winter cold soon invaded the territory again. Joseph didn't want to go through another winter with a sick family. In an attempt to keep the house warmer, he raked the leaves from the yard and placed them around the house's foundation. He didn't have enough leaves to bank the entire base of the house, so he collected bushel baskets filled with leaves from the woods. The bank of leaves was eventually more than two feet high, covering the whole foundation to just above the siding. His efforts paid off to some degree. The house stayed very warm, almost too warm. Joseph often found himself falling asleep early in the evening. Despite his efforts, snow and temperatures twenty-seven degrees below zero necessitated putting a heating stove in the basement. The leaves were just not enough insulation. When it was time to get rid of the dry leaves in the spring, Joseph started to burn them. A small gust of wind blew a handful of embers toward the house, igniting some of the remaining dry leaves. The old, wooden house immedi-

---

[85] Her mother and father (Fritz Franz) had moved to Chicago.

ately caught fire. Joseph never moved so fast in his life. He managed to get enough water from the garden hose on the house to put out the fire. Then he turned the hose on the burning pile of leaves. Almost as fast as it had started, the fire was over. That was the last time he used dry leaves as insulation.

In order to prepare for more complicated electrical work, Joseph began studying algebra, trigonometry, solid geometry, and calculus in the evenings. He eagerly read anything on various scientific subjects and attended every scientific lecture that was offered in the area. For example, one lecture was a discussion of Robert E. Perry's recent expedition to the North Pole. One of the most famous electrical scientists in the nineteenth and early twentieth century was Dr. Charles Proteus Steinmetz. When he lectured in Pittsfield, Joseph went with Elmer Newton, along with Elmer's girlfriend and his mother, Mrs. T. J. Newton, to hear him speak. It was a most exhilarating experience. At subsequent American Institute of Electrical Engineers (AIEE) meetings, he was fortunate to actually meet the great man. As inspiration, Dr. Steinmetz's picture hung in Joseph's office for the rest of his life. Joseph's acceptance as an associate member of AIEE, an organization Joseph always enjoyed, made him feel he had earned his status as a professional engineer.

While learning about new advances in science, Joseph also remained involved in the joys of the changing seasons and family life. Snow arrived again for Thanksgiving. An extraordinarily bright, full moon lit the landscape with its silvery beams. Joseph felt the scene was indeed one that inspires poetry. That weekend, Joseph made the first of many snow sculptures. The big snowman delighted the children who hated to see it melt. Russell had begun to say a few words and called the snowman "Da-da." Christmas was a little more bountiful because Joseph had saved enough money to buy a few extras. On Christmas Eve, everyone helped trim the tree. Russell even put on a couple ornaments. On Christmas day, the children were excited to see the presents under the tree. Natalie received a wagon and a doll. Russell received a teddy bear, lion, blocks, and an automated little horse.

Professionally, the year ended on a positive note. The Electrical World accepted two articles Joseph wrote in 1909. One was on how to put a pilot light on an electric motor in a coal elevator[82] to reduce the electric bill. The other was "Housatonic River Hydroelectric Plants,"[83] which included photos. He also started giving lectures on various electrical subjects. For example, at the local high school and town hall, he demonstrated how the telegraph worked.

His home life was very active by now. In early January 1910, Russell was running everywhere. He didn't have any trouble keeping up with his sister. In February, he could climb into a chair and get down again. When there was snow, the children enjoyed family sleigh rides and sliding down small hills on their own small sleds. Natalie was an incessant talker, and Russell smiled a lot. However, that winter was not kind to Natalia. She had difficulty sleeping and bad asthma attacks. She even took medicine for some sort of liver disease. To help her during the day, Joseph hired Annie Dorin to look after the children.

Family health issues created a need for higher pay. For three years with the Rogers Company, Joseph had put in long hours, including many nights. He felt he should be compensated for his time. However, there wasn't any sign of more money or a promotion. He finally asked for a raise. The company answered, "We'll see." The "when" was never answered. He began applying to other companies. In 1910, Rogers Electric was sold. Mr. Rogers retired in April. Mr. T. J. Newton took over the company and renamed it the Lenox Electric Company. Joseph knew there wasn't any chance for advancement with that company because the assistant manager, his son, Elmer, was clearly the choice to succeed his father. Joseph was given more responsibilities which his compensation did not match.

On February 11, 1910, Joseph received a phone call from the Fall River Electric Light and Power Company for an interview. As was his nature to take advantage of a business trip to enable a family visit, Joseph took the early Sunday morning train to Boston. He was anxious to see his brother, Eddy, who had recently married an Irish girl named Catherine Coyle. Joseph wanted to meet her and see how they were doing. He didn't get an answer when he tried calling, but he decided to go anyway. They had returned from church when he reached their Dorchester home.

The first impression was very distressing. They were living in a small apartment in a run-down building. Catherine was pregnant. Eddy was wearing a torn shirt, and dirty dishes were in the sink. Eddy was still working in the wholesale market for very little money. That impression lasted a long time. Joseph promised to come back after dinner and spend the night, even though he really didn't want to stay there. After lunch, he walked around Boston and had dinner before getting back at nine o'clock. Joseph shared the bed with Eddy while Catherine slept on an old sofa. The two brothers left at six o'clock in the morning. Joseph took Eddy for breakfast in a restaurant. Afterward, Eddy went to work at the market. Joseph caught the train to Providence and then took a trolley to Fall River. The interview with Mr. Hicks, an engineer, lasted until three o'clock in the afternoon, but it didn't lead to a job. He finally returned home at midnight. During a later one-day trip to Boston for another interview, Joseph saw Eddy again and met his mother-in-law, who was trying to help her daughter clean the apartment. Joseph asked Eddy to come to Lenox. Eddy said he would come on June 4. Joseph went to meet his train in Pittsfield, but Eddy never appeared. Joseph could only guess that Eddy never had any money, but they kept having babies.

In March, Natalia got away for a short break from the children and winter illness. She stayed in New York City for eleven days. She visited friends, saw a couple shows, and, of course, shopped. She came home a much happier person.

Joseph took care of the children in the evenings. The hired girl, Annie, took over during the day. She sometimes stayed over if Joseph had to work late or go to an evening meeting.

Shortly after that, Joseph took another overnight trip to New York City for several job interviews. Responses from the Livingston Company and American Conduit Company were encouraging. However, the Safety Insulated Wire and Cable Company was the most promising. The two articles Joseph had written for *Electrical World* had come to their attention. Joseph was flattered that they had been impressed with his writings. Mr. Condour was the headman in their New York City office at 114 Liberty Street. Their factory was in Bayone, New Jersey. The company had started a construction department, and they were looking for an engineer to take charge. When they offered Joseph a considerable increase in salary, he accepted without having to think twice.

While in New York City, he went to see if *Electrical World* would publish an article on the work of Samuel Morse, the man who invented the telegraph. Joseph had borrowed some of the inventor's papers from his son, E. L. Morse. The magazine was interested if he could leave the papers with them. He did, and they published an article. Joseph didn't get the papers back until a year later. At which time, he apologetically returned them to Mr. Morse.

Once again, he stayed with George and Carrie Dalhaus. In between interviews, he managed to see his sister, Wetty, and cousin, Valeria Franz Ullman, who now had a small daughter. He also quickly visited the Johnsons and Mrs. Carlson before returning to Lenox.

Natalia was still having a lot of discomfort, and she was easily depressed. It was beginning to impact the whole family. Because Joseph knew he might soon have to move again, he suggested she might like to spend a couple months with her family in Sweden. That seemed to cheer her up, and she began making plans for the trip home with the children. Together, they packed the dishes, carpets, and furniture. She bought new clothes for herself and the children as well as the tickets to Sweden and back. Joseph arranged to rent a space from Mrs. White to store their household effects.

At the end of June, 1910, Joseph resigned from the Lenox Lighting Company and the Stockbridge Lighting Company. He was tired of waiting for an increase in pay, even after long talks with Mr. Rogers. The offer in New York City was too good to turn down. Mr. Hull never believed he would actually leave.

On July 6, the family left Lenox for New York City. They stayed overnight with the Johnsons. The next day, Natalia and the children boarded their ship in Hoboken. The children hung onto their father and wouldn't let him go until they had seen every inch of the ship. It was sad for Joseph to see the ship pull out of the dock. For a little while, he wished he was also on the boat bound for a long vacation. However, reality set in. He soon had to start work again.

His boat ride was the ferry back to Manhattan for an appointment with the Safety Insulated Wire and Cable Company. They hired him as a chief engineer with a salary that greatly exceeded what he had been getting in the Berkshires. He was to begin on July 15.

CHAPTER TEN

Safety Insulated Wire and Cable Company

On July 15, Joseph arrived in New York City and checked in at the office of his new company. His first assignment would be to supervise the installation of conduits and cables in Middletown, Ohio. The company gave him seventy-five dollars for his expenses, plus a railroad ticket for the next day. After dinner with Luise, he spent the night with George and Carrie and left New York City early on Sunday morning. He grabbed a quick breakfast in Pennsylvania Station on his way to New Jersey to visit the factory. He wanted to become familiar with their available materials. After lunch in Jersey City, Joseph boarded the 2:14 PM train to Ohio.

Joseph traveled to Ohio in a compartment on a Pullman car. He had never traveled by train in such luxurious accommodations. The dining car was another treat. For dinner, he thoroughly enjoyed lamb chops, coffee, and ice cream for one dollar, including tip. After a sound sleep, he was up early. He was just in time to have a large breakfast of steak, eggs, and coffee.

The train arrived in Middletown, Ohio at 9:45 AM. Joseph was in the office by 10:17 AM, where he began work as a welcome visitor. His first chore was finding a place to stay. He chose one centrally located, run by Mr. and Mrs. Newberry, a pleasant couple with whom he felt an immediate rapport .

Over the first few days, Joseph was busy reviewing the contract for an underground cable system in the center of the city. He began ordering the necessary supplies by telephone to speed up the process. Mr. Rodier, the construction supervisor, came out from New York City to verify the order. Until the supplies arrived, there was really nothing to do. Joseph walked around the city, but he didn't find anything of interest, except seeing the circus that was in town.

When the supplies finally arrived in about five days, Joseph hired a union foreman and eight men. The foreman measured the line, and the men dug the trenches. Work generally progressed well, except on the few days when it rained. The men began laying the conduits as soon as the trenches were far enough along. Photographs were taken at various stages, so there would be a visual as well as a written record of the project.

Figure 16—Joseph on the construction site in Middletown, Ohio

Another part of his job was keeping the accounts. The New York Company opened a bank account in Middletown so Joseph could pay the men each week. He also recorded the types of supplies used and their cost.

Since arriving in Middletown, Joseph had not heard from Natalia for three weeks. It took quite a while for personal mail to catch up with him. Her letters eventually began coming every few days. She and the children had a good voyage over to Sweden. They were enjoying their new surroundings in Småland. Natalia sounded happy while visiting friends and relatives. She was not complaining of feeling ill at all.

Joseph shared interesting conversations with several people in the boardinghouse. Mr. Kinniside was with the telephone company. He and Joseph had long discussions about the telephone and the future of that field. Mr. and Mrs.

Newberry were also intelligent, personable folks. With them, Joseph went to the theatre, traveled to the airport several times, or just played cards. Another man, Dr. Sharky, had an automobile. He delighted in driving Joseph around the countryside on Sundays or in the evenings. Additionally, Joseph took long walks with a Mr. Fresch, which generally ended at the saloon in the city hotel.

Several problems arose at work when they began pulling cable. Joseph was informed that the job had to be finished as soon as possible because another assignment was waiting. Trying to rush the work proved counterproductive. Cable supplies ran out before new ones were delivered. The men were tired and showed signs of stress. They began making mistakes when joining the cables. Hot lead was splattered on them or others. Several people were horribly burned. Joseph was particularly friendly with one cable binder, J. O'Malley. He let O'Malley know he was doing a great job and he appreciated the work of all the men.

Suddenly, the most horrible thing happened. O'Malley was joining cables below ground level. Without any warning, an explosion occurred in the manhole. Rocks flew through the air. O'Malley was very badly burned. Joseph helped haul him out of the hole and immediately took him home in the construction wagon. That evening, he went to see his friend and stayed until ten o'clock. Joseph saw him again the next morning. It was hard to imagine the agony the man was suffering. No hospital was equipped to handle burn cases in those days. The best Joseph could do was take him to the Big Four Power Station for medical attention because they had a large first aid supply. Unfortunately, O'Malley did not survive his injuries. The vision of his blackened skin and distorted face haunted Joseph for a long time. He felt a great personal loss, but he had to put his feelings aside to finish the job.

When the last of the cable was in place, Joseph made the final connections and attached the cable to the overhead pole line. The cable was tested, and it worked. Two natural gas power plants, Rock Eagle and Big Four, were put on line. They were up and running before Joseph left.

On October 16, Joseph boarded the 6:00 PM train to Albany, where he spent the night. The next morning, he boarded the early train for Pittsfield. From there, he went to Lenox to see Mr. Rogers. He also checked on his things in storage and visited a couple friends before catching the evening train to New York City. The Dalhaus family put Joseph up again for the next few days. Natalia and the children were due back from Sweden on the October 19. Joseph went to Hoboken to meet them and waited until seven o'clock in the evening, but their ship did not arrive. That evening, he saw friends and went to an electrical exhibit.

In the morning, he was back at the New Jersey pier. The ship had docked just before dawn. A storm at sea, which slowed the ship's progress, caused the delay of the ship getting into port. It took a while to find the family in the crowd of passengers still going through customs. At last, Joseph and his family were reunited. Natalie and the children were exhausted from the long hours of standing in lines for their luggage and customs inspections.

Joseph took the family to the Dalhaus apartment and then left for the office. When he returned at 10:00 PM, they crowded together for the night. There wasn't much room for so many guests and all the luggage.

On Sunday, George and Joseph searched for a place to stay. Two days later, they found one room. However, when Natalia saw it, she didn't like it at all. In desperation, they finally had to take that place, just to have a room to themselves. On October 25, they found a furnished flat at 1175 Fox Street in the Bronx and moved the following morning. It was near an elevated trolley line, not far from Oak Point. After work, George helped move Natalia's big trunk to the new place.

Until the end of the year, Joseph generally worked in Mr. Condour's office. Except for one week in Easton, Pennsylvania, where he repaired some broken arc lamps, he made out the payrolls and did mechanical drawings for various jobs the company had.

Figure 17—Joseph in the office Safety Insulated Wire and Cable Company, New York City

In 1911, Joseph divided his time between drawing plans for special projects, designing electrical systems, and supervising construction sites. In all instances, the Safety Insulated Wire and Cable Company were the contractors.[86] Some of the design projects were for Johnstown, Pennsylvania; East Chicago, Indiana; Winnipeg and Indian Harbor, Labrador, Canada. Some of the special projects were overseeing the making of the first model of a concrete conduit and an ornamental lighting plan in Clinton, Iowa.

During this time, a frightening event took place. The sound of a huge explosion seemed to shake the whole building. Everyone in the office ran to Battery Park to see what it was, but nothing was visible. Via the evening newspapers, Joseph learned a ship loaded with dynamite had blown up in the harbor, killing many of the crew.

---

[86] See article in *Electrical World*, January 5, 1911 (with J. Franz photo).

During the early spring, Joseph traveled around, surveying sites for underground distribution systems in Passaic, New Jersey; Westfield, Massachusetts; and Easton, Pennsylvania. Most of the company's construction work was in Brooklyn or Manhattan. On Blackwell's Island[87] in New York City, Joseph had full charge of installing both conduits and cables. This was a particularly difficult job because the trenches in many places had to be cut through granite rock.

While the family lived in New York City, Natalia and the children, particularly Russell, were sick with one thing or another. They spent a lot of time at the doctor's office, getting medicine, and staying in bed. Additionally, someone broke into the apartment through an open window one night. Joseph reacted fast enough to catch him by the right arm, but the intruder was able to pull free and escaped. At least, the thief wasn't able to take anything that time. Joseph swore that was when he found his first gray hairs.

Nevertheless, there were good times as well. They entertained relatives and friends, and the relatives and friends entertained them as well. Their family life was fairly happy. New York City offered wonderful outdoor opportunities for the children in the Central Park Menagerie, Battery Park, Coney Island, Steeplechase Park, the aquarium, and several parks in the Bronx. St. Mary's Park was close enough to the apartment so Natalia and the children could walk to it during the day.

Joseph's brother-in-law, Karl Gumpinger, was unemployed for several months, so Joseph tried to have Wetty and their family over for dinners or picnics in a park as often as possible. Karl eventually found work in the Tiffany factory in Astoria, chasing silver and brass ornaments. He was an excellent craftsman and did really beautiful work.

The monthly AIEE meetings were always a pleasure for Joseph. At times, he invited Karl Gumpinger, George Dalhaus, or one of his other friends to go with him. A sampling of subjects discussed included aeronautics, construction of the Panama Canal, evolution of the world, and the cost of water. Thomas Edison gave one of the most exciting lectures. Joseph actually met and talked with Mr. Edison afterward. His wisdom and gentility impressed Joseph.

While Joseph was away from Stockbridge, he apparently was not forgotten. Whenever Charley Hull happened to be in New York City, on business at the Williams Coal Company or one of the other coal dealers, he invariably contacted

---

[87] Now Roosevelt Island

Joseph and extended a dinner invitation to one of his favorite haunts, mostly down by the fish market. Once they went to the theatre and saw *Excuse Me*.

On these occasions, Charley would often ask, "Wouldn't you like to come back to Stockbridge?" Unfortunately, he never mentioned a word about returning at a suitable salary. Mr. B. H. Rogers also came down once. He took Joseph to a nice restaurant for dinner and the theatre to see a zany review called *Got Rich Quick*. However, there wasn't an offer of a job possibility from him either.

When the town of Lee asked to get power from the Stockbridge Lighting Company, having been refused power from Lenox, Joseph was asked to assay the Lee situation. Lee only had electric service at night from a 133-cycle steam plant. During the day, the Lee Electric Company served the Lee Lime Company, White Terrazzo and Marble Company, Standard Lime Company, and Lee Wire Works. All of these companies needed more power. On two weekends, Joseph went up to Stockbridge as a guest of Charley Hull to make a proposal to lay out a nine-mile line. He figured the cost of the installation to be $9,000, but it didn't include anything for his own time. The proposal was not accepted.

Joseph's last job for the Safety Insulated Wire and Cable Company was in Utica, New York. As in Middletown, Ohio, he had to move to Utica to oversee the work. He checked in with the chief engineer at the city hall to hire the personnel he needed. On Sunday, Joseph enjoyed the luxury of a fine hotel, including good food, great coffee, and a good bed. Around July 1, he found a furnished apartment on South Street and sent for his family. When they arrived on the Fourth of July, Natalia thought they should be more centrally located. They found another place on Lafayette Street that was more central, but, after one night, Natalia didn't like the furnishings, décor, or surrounding area. They returned to the South Street house, which suited her better, and rented it. He at least had a pleasant place to go at the end of the day.

Utica had a very powerful union. Joseph, who represented an outside contractor, had a lot of trouble finding men who would work for him. On July 19, the union boss, Frank Camello, confronted him and said Joseph had to hire his men for the job. Joseph was not one to be bullied into anything. Joseph spoke to Commissioner Beebe about the incident. When he took the men he hired to unload the supply wagons, he had another confrontation in the park with Camello and some of his men. Joseph's people were intimidated and backed away. For several days, he had to face the threats and taunts from the union bully. Each time this happened, one or two men would quit. Joseph then had to hire new men. When he finally did get a foreman and a crew, the work progressed very slowly. Mr. Rodier arrived from New York City to help solve with the problem. This action only made the men he was able to hire less cooperative and unfriendly. Joseph really felt frustrated and forlorn. It was the first time an angry mob had condemned him as a "damned foreigner" unworthy of overseeing "good American laborers." He had worked with people who were jealous of his skills and ability to please his bosses before, but this situation was truly serious and prejudicial. This was the first time that Joseph had absolutely no rapport with any of his crew.

In contrast to the work situation, he and his family did enjoy some things that summer. Over two weekends that summer, they took special excursions to see some of the country in upper New York state. The first was to the Thousand Islands Park in the St. Lawrence River. They took a train to Clayton. Then they took a boat to Alexandria Bay, where they had a lovely two-hour boat ride through many of the islands. The next trip was to the Adirondack Mountains and another boat ride to Arrowhead on Big Moose Lake. The mountainous scenery on both trips was gorgeous. They saw breathtaking views of the heavily forested wilderness. Sharing these trips with his family helped Joseph overcome some of the distress and degradation he faced on the job.

Charley Hull asked Joseph to attend the August board of directors meeting of the Stockbridge Lighting Company. He had submitted the plans for the transmission line to Lee. Once again, Joseph was invited to stay with Charley and his wife. The evening Joseph arrived, Charley treated Joseph to a lovely dinner in the Red Lion Inn. The next morning, when he went into the sitting room to wait for Charley, he surprised Mrs. Hull, who was still in her nightclothes. In embarrassment, both quickly left the room. Joseph returned to his bedroom to wait for Charley's knock on the door.

At the meeting, the board voted to borrow the great sum of $9,000, sign contracts with Lee Electric and the various other Lee companies, and proceed with the work. When a stockholders' meeting was held later that month, Joseph traveled to Stockbridge again. This time, he stayed with the Rogers family in Lenox. At the meeting, he presented the plans, complete with all of the figures and projected income from this line. Mr. P. P. Bowker, a past president of New York Edison Company, made a motion to authorize the directors to borrow the money and proceed with the work. The Honorable Joseph H. Choate, another one of the stockholders, was not in favor of going into such an undertaking.

Choate said, "Gentlemen, I thought we formed this company as a neighborly affair to help one another to obtain electric lights for ourselves, not to go into a large corporation and spread beyond our town limits."

Mr. Bowker immediately withdrew his motion. Such were the acts of gentlemen among gentlemen. It delayed building the line for two years. It might be called ultraconservatism, but it was a disadvantage for the small companies.

Joseph returned to Utica and faced more conflicts. This time, his fight was with the dishonest inspector and city engineer. Again, he had to call Rodier in New York City. Rodier told him to come down to the New York City and he would try to help.

Right when Joseph returned to the city, Charley Hull just happened, or perhaps by intention, to also arrive in New York City. Charley had somehow won with his board of directors, and he brought a firm offer for Joseph to return to the Stockbridge Lighting Company at an increased salary. Mr. Westinghouse had decided to give up his private power plant, so Lenox now also wanted power from Stockbridge. The line to Lee[88] was also being built. Joseph would be their power engineer.

Neither Mr. Rodier nor Mr. Condour wanted to lose Joseph. They had several large construction jobs and Utica job to be completed, but they wouldn't match the higher salary that Joseph had been offered. This was a good company, and Joseph was reluctant to leave, particularly since he hadn't been able to finish in Utica. However, they parted on amicable terms, and he stayed in touch with them. Over the years, he often ordered supplies for various projects from them.

The additional money was certainly welcome, but Joseph had other motivations to take the job. Being married with two small children, Joseph no longer enjoyed traveling as much. It wasn't any fun for the family to be relocated all the time. Life in a furnished apartment was fine as temporary housing, but they were eager to have their own things again. Joseph accepted the offer to return to the Berkshires with the stipulation that, in addition to the salary increase, he asked for the privilege of doing any and all other work as his own private enterprise. Mr. Hull agreed, just as long as he kept the plant going and took care of all the work of the Stockbridge Lighting Company.

---

[88] Eventually, Lee went to the Western Massachusetts Electric Company, which had absorbed the Pittsfield Electric Company, for their power.

Chapter Eleven

Settling in the Berkshires

The family remained in Utica, New York, until Joseph found housing in Berkshire County, Massachusetts. The new job was time-consuming, and he could only come to Utica over a couple weekends. At the end of August, he found a house in Lee and moved their furnishings from Lenox. The family finally arrived. There was a lot to do before Natalia was satisfied with her new home. Some new furniture was ordered, but it arrived broken or obviously repaired. It had to be sent back. Windows were washed, drapes were made, and painting was done. The house had a big backyard in which the children could play. As a surprise for Russell's birthday in September, Joseph made a wonderful seesaw that the children enjoyed for several years.

With colder weather, illness again plagued the whole family. Russell and his mother suffered the most. They called Dr. Stockwell many times that fall and winter of 1911–1912. He prescribed one medicine or another, but nothing seemed to help for long. Adding more complications, Natalia was pregnant again.

As power engineer for the Stockbridge Lighting Company, Joseph was busily reading electric meters each month, checking on the power stations to make sure they were running properly, drawing up plans for new town lines, and calculating the increased income from new customers.

As a private consulting engineer, he laid out transmission lines for the Lenox Lighting Company. For local firms, such as the Lee Lime Company, Granger Lime Company, Tobey Lime Company, Miller Lime Company, and White Terrazzo and Marble Company, Joseph designed the necessary buildings, electrical substations, and power lines. In many cases, he recommended and supervised installation of the electrical equipment as well. Under a commission contract with General Electric Company, he began selling equipment and appliances. For the Stockbridge Water Company, he laid out and installed the water pumping plant and installed the sewer pumping plant. In addition, he prepared plans and specifications for the Glendale water supply.

Joseph had worked for B. H. Rogers in Lenox in 1906, and they remained friends over the years. Or so, that is what Joseph thought. Mr. Rogers was originally from Hingham, Massachusetts. He was of very old Yankee stock. He prided himself on his great ancestors. He was a teetotaler and always preached about honesty and piety. He and his family had always treated the Franz family very kindly. In October 1912, Mr. Rogers invited Joseph and Natalia for dinner. They enjoyed a good New England meal. There weren't any drinks except for coffee and tomato juice. After dinner, Joseph was motioned to one side while the wives chatted elsewhere.

"I say, Joe," Mr. Rogers started cautiously, "I have a great many good accounts on my books, but I cannot press these people for money. They have plenty! Nevertheless, it would insult them if I asked them to pay their bills. I would lose their work. They are all good clients, and they just don't realize that a businessman needs his money to pay his bills. I need some money now to purchase materials for the Fahnstock job. Therefore, I went to the Lenox National Bank and asked if they would loan me the money on my listed accounts. They said they could not loan anything on book accounts, but, if I got an endorser on a note, they would lend me the money. I told them I never had anyone endorse a note for me and asked whom they would suggest for an endorser. They suggested you!"

Joseph was somewhat overcome to think a bank, where he only had a saving account, would suggest him to endorse a note. Mr. Rogers's note was for $3,000. Joseph was assured it was only for three months because the money owed to him would surely come in by then. Joseph signed. Three months passed. Another note came for three more months. Six months passed. Again, Joseph's signature was needed. It continued. After a year, Joseph went to see Mr. Rogers. He reminded him that this was to be only for three months and he didn't want to sign anymore notes.

"Then," Mr. Rogers declared, "If you don't want to sign it, you can pay it."

There wasn't any alternative. After years of pleading and begging, the note was eventually paid back in small increments. That is what you call "Yankee Honesty." It was harsh lesson. Joseph never signed another note and never loaned anymore money. One learns by bitter experience.

In his rare spare time, he preferred to be ice-skating and sleigh riding in the winter. In the summers, he'd swim, boat, and garden. He'd walk in the woods at any time. Joseph was never comfortable sitting around and making small talk with people, but the AIEE meetings were always stimulating because they offered new knowledge and food for introspection. He appreciated the opportunity to meet fellow scientists. He felt an intellectual affinity with them.

Early in 1912, he met Dr. W. McBurney from Boston, who had his own experimental laboratory on his Cherry Hill Farm in Stockbridge. Joseph was invited to the laboratory many times. The two men experimented with tungsten and carbon arc lamps, stationary cylinders, and other possible uses of electricity. They even discussed the possibility of setting up a technical high school where they could teach. Unfortunately, their dreams ended in 1913 because of Dr. McBurney's sudden death.

On May 31, 1913, the Franz Family entered the automotive age. Joseph had been looking at cars for some time and finally decided to buy one for the family. It would also get him around the county more easily than to depend on public transportation or horseback. He went to Pittsfield to purchase a Chrysler touring car from Sisson's Auto Shop. He took a few lessons there and drove the car home. That day, no one could have been more proud and pleased with himself than Joseph. He could tinker with a new toy. Like most autos then, it constantly needed tuning up or repair.

Figure 18—Joseph's first car, a Chrysler touring car

Once again, home life was traumatic. Natalia's pregnancy was extremely difficult, and she was ill much of the whole nine months. By mid-June, she really couldn't care for the house and children alone. They hired Miss Potts as help, but it was to no avail. The pregnancy terminated with a stillbirth on July 7. Miss Potts stayed on for another two-and-a-half weeks, caring for the family until Natalia was well again. To help his wife recover from the loss of their child, he planned a big event for just the two of them, a delayed honeymoon. Their Lenox friends, the Jerkins family, agreed to take care of the children while they were gone.

It was a five-day bus tour to the Adirondacks and Vermont, beginning in Pittsfield. The bus first drove to Lake George, where nature was displayed in spectacular mountainous views around the pristine lake. Joseph always loved rural areas, untouched by human invasion. They stopped for the night in Port William Henry and had a good dinner. The next day was bright and sunny, so Lake Champlain sparkled as a ferry took them across to the Vermont ship landing. Another bus met the group and drove through the verdant Green Mountains to Burlington. They stayed at the Burlington Hotel and took side trips to Shelborn and the historic S. Webb estate. On the fourth day, they went to the Bluffs Point Resort Hotel on Lake Champlain, one of the grand summer resorts of the time. The imposing structure faced the lake, so most of the rooms had an excellent view. Dinner was also excellent. Later, they listened to fine musical entertainment. Afterward, they enjoyed a ninety-minute moonlight boat ride across the lake and back, a treat Natalia and Joseph agreed was the highlight of the trip. On the fifth day, they picked up the children on their return home.

In the fall of 1913, little Natalie started kindergarten in Lee. Russell seemed to have outgrown most of his early illnesses, and Natalia's stomach problems were diagnosed as gallbladder attacks. The proper medication helped her finally get relief.

In November, Joseph's vote helped elect Woodrow Wilson as the twenty-eighth president of the United States. During his tenure, he introduced the graduated income tax[89] and established the Federal Reserve System. The Balkan War in Europe increased the demand for American raw materials and finished goods, so the country was experiencing a general prosperity. In New York City, both the Woolworth Building and the new Grand Central Station opened. Many other new buildings were pushing skyward as well.

---

[89] Ratified in 1913, this was included as the sixteenth amendment of the United States Constitution.

With the assassination of Archduke Franz Ferdinand and his wife in 1914, the Great War[90] started. The Austro-Hungarian Empire declared war on Serbia. Germany declared war on Russia and France.

Yet, in other parts of the world, life continued more placidly. The Panama Canal opened, and ships could now cross to the Pacific Ocean without the long, hazardous journey around the horn of South America. E. H. Shackelton led a daring expedition to Antarctica. Joseph followed his journey in later reports. Everyone's miraculous survival awed him.

Joseph was making good money, and his work seemed reasonably secure. Natalia and Joseph liked living in the beautiful Berkshire Hills, and both had made several friends who had children that were the same age as theirs. Natalia was still having health problems, but, in the country surroundings, she seemed happier than in the city. Because Stockbridge was the place of Joseph's principle work, it seemed a natural location for them to have their permanent residence.

They found a piece of property on Elm Street with a small house on it.

Figure 19—The small house on the Elm Street property that Joseph bought

---

[90] World War I—it was not called by this name until the start of World War II

At the time, Joseph had exactly $3,000, his and Natalia's entire fortune. It was just enough for the down payment, so Joseph completed the title search himself. A Dwight family who currently lived in the Philippines had once owned the property. They had already signed off on the land, and there weren't any other liens or restrictions. With that information, Joseph bought the Elm Street property, free and clear, from Captain Miller for $4,500.

At the same time, Dr. Henry C. Haven, a wealthy widower with an extensive estate on Ice Glen Road, offered Joseph ten shares of Stockbridge Lighting Company stock for $1,000. Even though Joseph had just made a deposit of $3,000 on the Elm Street lot, he agreed to the offer and wrote a check for $1,000.

He presented the stock certificate for $1,000 to the Housatonic National Bank on Main Street and wanted to borrow $1,000 at six percent. The stock paid six percent in dividends. Mr. Seymour,[91] the president of the bank, hedged on his offer. Joseph had already found out this was customary with Yankee businessman. When Joseph opened an account in early 1914 for $1,000, Mr. Seymour had said, "So, you are coming to Stockbridge to live. Well, I am sorry for you."[92]

This time, he tried putting Joseph off.

"Mmm! I don't know about this. Come in this afternoon." When Joseph came back, he responded, "I would rather not loan you anything on this stock. I talked to Mr. Hull. He says the stuff is not marketable and has no value."

"Not marketable and had no value!" Joseph thought.

Joseph knew Mr. Hull was already dickering with the C. D. Parker Company to sell at a price of $175 per share. Apparently, that was how he had acquired a considerable amount of stock from people who sold estates at a ridiculously low figure. This was apparently the same practice if anyone wanted to sell some of the local water company stock. Joseph was worried. He had written a check for $1,000, dated a day ahead, without any funds. In desperation, he went to the Lenox National Bank and could borrow the full amount without any questions asked.

---

[91] Mr. Seymour was the son of the man who established Seymour General Store, currently Williams Store.

[92] This was an implication that Joseph was not welcome in the community.

Adding to the insult, when Charley Hull recorded the transfer of stocks on the books of the company, he saw that Joseph now owned some stock. Accusingly, he said, "Say, Joe, where did you get a hold of the ten shares of Stockbridge Lighting Company stock?"

After Joseph told him the story of Dr. Haven, Charley cajoled, "Now Joe, you don't want this stock. I'd kind of like to have it for Charlotte[93]."

"Well," Joseph countered, "if it would be good for Charlotte, it ought to be just as good for Natalie." He then kept the stock.

This did not sit too well with his Yankee boss. Six months later, Joseph sold the stock to C. D. Parker Co at $175 per share, a privilege "no damn foreigner should enjoy." Apparently, only the wealthy American-born were supposed to be able to invest in the stock market.

Meanwhile, Joseph had drawn his own plans and specifications for a commercial building and his house. The Captain Miller house sat on the front of the lot. He wanted to move it to the rear of the lot so his family would have a place to live while the new buildings were being built. He had to borrow all of the money, so he wanted to get assurances from potential tenants. First, he received a letter from the New England Telephone Company stating they would occupy the second floor of the commercial building, subsequently known simply as the Block. He received cost estimates from the Clifford Building Company for both the house and the Block. Their figures were $12,000 for the block and $4,500 for the house. With these estimates, Joseph went to the banks. Stockbridge again refused him.[94] Lenox Savings Bank allowed him $5,000 on the block mortgage. Housatonic Cooperative Bank offered him $3,500 on the house. All were nonamortization mortgages, which was customary before the banks were badly hit in the Great Depression. During that time, many pieces of real estate were sold much below the value of the mortgage. Consequently, the actual cost for the Block was $4,500. The value of the house was $3,000. Joseph had money to spare, including to move the house and pay for his stock, in the bargain. The money was not fully paid until the buildings were completed. Because the mortgage was not amortized, it could be paid off at will. He paid off both mortgages in a matter of ten years.

---

[93] Charlotte was his adopted daughter.

[94] Joseph never did receive any financial help from this bank.

Neighbors tried very hard to stop Joseph from building anything. The Van Deusen family owned the hardware store just across Elm Street, and their house stood next door to it. They emphatically informed Joseph that the lot the house stood on was sold to the former owner, Captain Miller, by the Dwight family with the express proviso that nothing must be built there because of the proximity to the Stockbridge Public Library. The Dwight family had given the land for the library, but there was nothing in the deeds about any such restrictions. Joseph went ahead with the building. The "damned foreigner" was considered a sort of pirate for not respecting sacred promises that were supposed to have been made.

When he started building the Block, Joseph discovered his land extended three feet beyond the front line of the building next door, Rathbun's Market Building, which was then a fish market. At the time, Stockbridge did not have zoning setback requirements. The fish market had a raised platform in front of the building that was used to display fish. It extended the building to the property line. For aesthetic reasons, Joseph felt all buildings should be kept back off the street by the current three feet, which the fish market building established. He asked for an agreement that, if he kept his building back, his neighbor would also keep that building clear to the front wall. Mr. W. E. Rathbun did not want Joseph to build out, but he would not commit himself to give any legal paper. Joseph started to dig out to the property line. The forms for the foundation were in when a showdown occurred between Joseph and Mr. Rathbun. Charley Hull was asked to intervene. In the end, Joseph got a paper drawn up and recorded. However, it cost him a lot of money to change plans because he had to dig out further in the rear, take out and replace the forms, and fill in the hole at the front of the property. Subsequently, the town enacted a setback law.

That was not the end of the problems with Mr. Rathbun. The eaves on the house behind his market projected over Joseph's property line. Mr. Rathbun was told to remove the offending eaves.

Joseph asked, "What are you going to do about the eaves?"

"You know what you can damned well do!"

Rathbun's obscene gesture said the rest. Several witnesses saw this scene, so statements were taken from Louie Fenn, J. M. Buck, and Mrs. Pilling, who had also participated in the argument. Joseph went to Pittsfield and searched the W. E. Rathbun title. That afternoon, Joseph cut back the overhanging sections of the eaves and erected a wall of his building on his side of the property line. Mr. Rathbun became enraged and hired a surveyor to measure the line between the two properties as well as the street line. He only found out that Joseph had indeed measured correctly.

Through all this, the Block[95] was built with a twelve-foot concrete block wall and garage that extended from the back of the building to the end of Joseph's property. The telephone company moved into the second floor. The first tenant downstairs was Monroe's Bicycle Shop.

Figure 20—Monroe's Bicycle Shop in Joseph's Block

---

[95] Today, it is known as the Mercantile Block.

Subsequently, the Stockbridge Post Office took over most of the ground floor. The bicycle shop moved to the small retail space in the back that opened onto the drive.

When Joseph bought this property, two families lived in the house. They had toilets and sinks, but they didn't have any bathrooms. It was hardly luxurious accommodations, but Joseph planned to use this as temporary housing for his family. That way, they could save the cost of a few months rent until the new house was finished. At which point, the old house would have been torn down.

It was a fine idea, except for the impediment of the Pilling affair. One Mr. Wilson lived in the rear of the house. Mrs. William Pilling lived in the front part with her daughter, Mrs. Potter; teenage son, Eddy; and granddaughter, Gwen Potter. All of them were asked to vacate so Joseph could move in his own family. Mr. Wilson moved without any fuss. But the Pilling family was different. They, particularly the daughter and her child, then about six years old, were good friends of the Van Deusen family. Mark and Ed Van Deusen spent considerable time in company with the Pilling family. What happened during their lengthy visits is anybody's guess. There was a lot of drinking, as evidenced by the empty beer bottles thrown out in the rear. They were repeatedly asked to find some other place to live. Ed Van Deusen explained the Pilling family had lived there for more than thirty years.

"They had to live in a central locality because they made their living by taking in boarders. If they had to move, it would mean taking away their livelihood. Couldn't you at least move the front part of the house to the rear of the lot?"

Joseph foolishly softened under their pressure and agreed. He received estimates on moving, cutting in two dormers in the roof to make two bedrooms and a bathroom upstairs, a luxury they had never enjoyed. The rent was raised from ten dollars to fifteen dollars per month. Joseph installed a heating system. Steam was supplied from a central heating plant in the Block. The same steam plant would furnish heat for Joseph's house when it was finished.

Figure 21—Joseph's home on Elm Street in Stockbridge, Massachusetts

The Pilling family stayed in the house during the move. As soon as the house was on the skids, many defects in the old house were revealed and had to be corrected. The protruding nails left from the worn, soft wood floors were ruining the carpets. New hardwood floors solved that problem. Blackened walls and ceilings from the coal stoves meant complete redecoration. The heat was never satisfactory. The fire in the Block was banked at ten o'clock in the evening, but the Pilling family stayed up with their company until the wee hours of the morning.

The heating problem was finally resolved when Joseph installed an independent heating plant. Now all was fine. For once, they had all of the heat they wanted. Alas, the fuel bills were too high. After a few months, the steam heating was discontinued. The old potbelly stoves were reinstalled, causing marked up walls and ceilings. Joseph was supposed to redecorate every year.

Completely frustrated, he asked them again to move, but it was to no avail. The only recourse was to raise the rent. When it increased to twenty dollars per month, they finally moved out. The "damned foreigner" was the meanest man in town. The house was finally torn down. Joseph's attempt to satisfy his neighbors cost plenty, but he finally began to cash in on his investment.

Another Yankee trick that befell Joseph was the Hopkins Lumber Mill owing Charley Hull a considerable bill for coal. Joseph was asked to buy all of his lumber from Hopkins and then turn the payment for the bills over to Mr. Hull to balance out his account with Hopkins. After the first bill turned over in this manner, Hopkins naturally sold Joseph all of the junk lumber he had. Joseph didn't have any redress, but Charley's account was squared. There was no way out. His job depended on his compliance. In 1930, when he built the market, Joseph had to acquiesce to the same rotten deal.

Building the house was begun on March 22, 1915. Four months later, on July 27, the family moved into their new home. Very well built with hollow block tile, the house had a stucco exterior, lath and plaster interior, and a peaked, asbestos roof. The style is described as American Craftsman, giving Joseph credit as the designer. The house is symmetrically balanced. Windows on the second floor are directly over windows or doors on the first. On the first floor is a living room with a bay window facing north and corner fireplace. There is also a dining room with a corner fireplace, an entrance hall with a stairway to the second floor, a kitchen, and Joseph's office with its own bathroom and private entrance. The office is where Joseph set up his drawing board, conducted consulting work, and met with prospective clients. Pocket doors close off the archways between the hall and living room and between the living room and dining room. On the second floor are four bedrooms, two sleeping porches, and a bath. There is a full basement and attic. Across the front of the house stretches a porch that was once screened. After the first car was replaced, the leather-covered backseat of the old Chrysler hung from the porch ceiling and served as a swing for years.

The family settled into their new surroundings very well. Natalia had more room for entertaining her Swedish friends and the *kaffe klatches* were frequent occurrences. Joseph laid out a big garden in the backyard, complete with a stone birdbath and its own running water in the center.

News from the family in Europe was still coming, mostly at Christmas or Joseph's birthday. Sometimes it came via Wetty or Karl in New York City. In 1911, Joseph's father had bought a house in Gumpolskirchen, a small suburban town near Vienna. Rosa and Alma continued to live at home with their parents.

Joseph's mother died on October 22, 1915, her sixty-second birthday. Her last correspondence was a month before her death. She signed a card Rosa had written. Joseph was deeply saddened that he had not been back to see her again, but he was happy he went home at least once.

Ludwig and Pepi had their own apartment in Vienna. Ludwig had received notice that he would have to serve in the army again because of the war, but he had not yet been notified of a specific date. Pepi was looking for work because she knew things would be hard without his regular income.

The last correspondence from Europe before the war was to inform them that Joseph's father had remarried in 1916. His second wife, Adele, was only twenty-one years old. He was sixty-five. The family was scandalized. It hit his sisters, particularly those remaining in Europe, the hardest. Years later, the family in America learned that Rosa bought a small house, which she shared with Alma until Alma got married. Their father had immediately started another family. They eventually had two half-brothers, Karl and Otto. Joseph's older brother, Karl, was hurt to hear that his father gave his name to a second son, albeit he was the eldest of a new family. Joseph eventually understood the loneliness that was part of his father's motivation.

At the same time, Natalie started first grade. The school was just across Elm Street and east around the corner on Main Street. With her mother's temperament, she easily made friends. The children were unaware of the difficulties and prejudices their parents faced. Their parents tried not to pass on their problems.

Natalie and Gwen Potter walked to school together. They communicated with each other by sending notes over a small rope, strung like a clothesline, between the old and new houses. When Frances Pilling visited his grandmother, he climbed onto the roof of the garage and jumped into Natalie's waiting arms. Eddy, his father, grew up to become a plumber's apprentice. Later, he had his own plumbing business. Much later, Frances took over the plumbing business from his father.

Natalie also made friends with Esther Van Deusen, who lived across the street and was about her age. They did a lot together after school and on weekends. Natalie was over there many mornings when Laura, Ed's wife, would give her sugar cookies, molasses cookies, or whatever she had baked. The older folks never set foot in the Franz house. They were always jealous that Joseph had bought the property and built on it. Mark and Ed never seemed to do much work because they were constantly standing outside their store while watching the activities on the street.

Life in the larger world during this time was chaotic. During the teen years of the twentieth century, the Great War was raging in Europe. It wasn't until the *Lusitania* was sunk in 1915 and Eric Muenter planted a bomb in the United States Senate reception room did America feel any direct connection to the war. However, there weren't any great changes in American life. President Wilson married Mrs. Edith Galt, Booker T. Washington founded the Tuskegee Institute, Alexander Graham Bell in New York City made the first transcontinental telephone call to Thomas A. Watson in San Francisco, and Einstein postulated his theory of relativity.

Moreover, Joseph continued having his own battles to keep his career in focus.

CHAPTER TWELVE

Expansion

Consolidation in electric companies was occurring in 1915 and 1916. Joseph had to remain sharp or be lost in the shuffle. C. D. Parker and Company had already bought out Lenox and Great Barrington, so there was nothing else to do for the little Stockbridge Company except get on the bandwagon or be squeezed out. The sellout was accomplished in late 1914 without the transfer of one penny. It was called "high finance." All of the stockholders were offered $l75 for their $100 stock certificates, but no actual money was paid. A ten-day note at five percent was given. Before the note was due, one could get either cash or a new certificate for one paying six percent interest, which meant that twelve percent was paid on the original stock. The old Stockbridge Company had earned that dividend but only paid six percent. The balance of the income was spent to make extensions. Thus, the stock was well worth the $175 they offered for it. Two shares in the new company for one of the old were offered for sale at the bargain price of $185. Most everyone bought the stock back. This way, the C. D. Parker and Company took over the control of the Stockbridge Company. Mr. Bowen Tufts from Boston became vice president. Five more directors from Boston were added. Three, including Charley Hull, Allen T. Treadway, and Alexander Sedgwick, were from Stockbridge. Charley Hull was still president of the Stockbridge Electric Company.

    Richard V. Happel, a reporter for the *Berkshire Evening Eagle*, once described big business of this era as "a loaded crap game on an Oriental rug." The players often lost. In the Stockbridge Company sellout, the stockholders paid the C. D. Parker and Company ten dollars per share of stock to have the control taken away from them. Some $21,250 more capital than the original stock issue had been invested. It never had been increased because the large earnings paid for all of the extensions. C. D. Parker applied to the Electric Light and Gas Utilities Commission to increase the company's capital stock by that amount. The commission did not allow the capitalization. They refused because this was not C. D. Parker's capital because they did not put any new money into the company. They simply used money rightfully belonging to their customers to make the extensions. C. D. Parker and Company got out of it by simply carrying the $21,250 on the books as a premium on capital stock.

From the beginning, Joseph often wondered why Charley Hull was picked to be president. C. S. Mellon practically pushed this position in Charley's lap. Later the C. D. Parker outfit continued with him at the helm. In time, the reason became obvious.

In the early days of the company and in certain locations, there was trouble with pole type transformers breathing. Breathing was caused by the heating and cooling of the transformers. Moist, outside air entered the transformer, where it cooled and then condensed on the walls of the case and cover. The water dripped to the bottom, where it eventually reached the coils. Thus, the transformer burned out. The transformers were purchased from the Westinghouse Company. Because these transformers failed after one or two years of service, new ones were ordered to replace the ones that had burned out. The Westinghouse Company was expected to take back the old units and exchange them for the new ones. They did not see it that way, and a lot of correspondence happened back and forth.

Eventually, the Westinghouse Company sent their chief engineer, Robert Scott to settle this matter. He arrived with a Westinghouse sales representative to meet with Charley Hull and Joseph. The meeting was held in Charley's home office, which was furnished with the bare necessities for himself and his staff of two others. The tight space required the staff to retire to the kitchen. Robert Scott took the chair at the flattop desk. Joseph sat on the edge of the flattop desk, and Charley occupied the chair at his rolltop desk. This meant the Westinghouse salesman had to lean against the radiator in the corner near the door. Robert Scott led the discussion.

Charley, not being a technician, did not object to Joseph interrupting whenever he saw fit. This did not please Robert Scott at all.

Turning squarely to Mr. Hull, he said, "Now are you or is this man going to decide this question?"

Charley Hull, with his Yankee shrewdness, responded quickly, "I think we will sort of decide it between us."

Robert Scott was defeated. Stockbridge did not have to pay anything for the replacements. Further, all other transformers then in service were equipped with tight gaskets to prevent breathing.

Joseph had mixed feelings about his longtime friend. Charley Hull had the rare faculty to say the right thing at the right time. He was always on the fence. He seldom took any direct action himself. Instead, he always managed to get someone else to do it. As a result, he didn't make any enemies. Sometime, this quality backfired. For instance, when Dr. Austin Riggs started the Riggs Foundation sanitarium, there was much opposition to it. The cottage people especially felt it would change the town's character. C. S. Mellon led the opposition. The foundation came into being anyway. Charley had not taken sides. The Aymar house, which was next to Charley's house on Main Street,[96] was used as a boardinghouse for the patients. Charley complained about the habits of the patients, including their immodesty in disrobing without pulling the shades and so forth.

Mr. Mellon, who had become one of Joseph's good friends by then, sneered, "I've got no sympathy for Charley. He sat on the fence, and it fell over."

Charley Hull often said, "There are those who make the cannon balls and those who fire them." Charley never fired a single one.

Stockbridge did still hear of world events. In November 1917, Woodrow Wilson was reelected for a second term. The war was brought closer to home when saboteurs blew up a munitions arsenal in New Jersey. Still, the president was negotiating for a peaceful resolution. Despite President Wilson's efforts to convince the belligerent countries to settle their problems through diplomacy, the United States and Cuba both declared war against Germany in April 1917. War was declared against Austria and Hungary in October.

American males from ages sixteen to fifty began volunteering for the army, hoping to see action overseas. Many more servicemen were needed, so the first selective service conscription for men ages eighteen to thirty was put in place in June. In July, the first draft began. Joseph was thirty-five and too old for the first registration. By June 1918, when the second draft extended the age to forty-five, Joseph was considered an essential worker because of his work with the power industry. He was thus exempted from military service.

Whatever prejudice existed against him as a "damned foreigner" was further intensified with the war. He was now the enemy. He had placed a small sign in the window of his office that simply read, "Joseph Franz, Engineer." When people began throwing tomatoes and other stuff at the window, he took down the sign and never put it back.

---

[96] The Hull house is now incorporated into the Riggs property on Main Street.

Because of the draft, fewer men were available for maintenance of the system. Joseph had an extension phone installed in the bedroom, so he could receive emergency calls in the middle of the night. These would come during a bad thunderstorm or the worst snow or ice storms. He'd have to go out to help repair the downed wires. During the winter of 1918, a particularly bad ice storm hit around midnight. It brought down one of the high-tension lines. Joseph and two other men arrived on the site and saw the broken line bouncing all over the place. Even in the dark, one couldn't miss it. Every time the line hit the ground, it shot off sparks, lighting the night like fireworks on the Fourth of July. One man climbed the pole to start the repair. A new man on the ground with Joseph was unaware of the danger. The wire bounced close to where he was standing.

The man on the pole yelled, "Watch out for the hot wire!" As soon as he said that, it bounced again.

"You mean this one?" asked the new guy, grabbing the wire to stop it from bouncing around.

He was immediately grounded with 2,200 volts of electricity coursing through his body. Luckily, Joseph immediately knew what to do. He found a log and some dry newspaper that he wrapped around the log. He raced to where the man was shaking like a leaf. Using the insulated log, Joseph knocked him free of the wire in a swift blow. He had a hole in his heel the size of quarter, but he lived to tell the story!

In recognition, Joseph received the following letter from Henry Parsons, one of the directors of the Stockbridge Lighting Company.

```
Palmer, Mass.
April 11, 1918
Mr. Joseph Franz
Stockbridge Lighting Company
Stockbridge, Massachusetts

Dear Sir:

I was in Great Barrington on Tuesday, April 9,
and regret very much that I was unable to meet
you and personally congratulate you upon the
presence of mind and the knowledge you dis-
played when you successfully brought Mr. Murphy
```

```
back to the land of the living. Mr. Parrish
outlined to me just the conditions. I concluded
from what he has told me that, had you not been
nearby and been able to think promptly and
properly, the man would not be alive today. You
are certainly to be commended in this matter.

With kindest regards,
I am very truly yours,
Stockbridge Lighting Company
(signed) Henry Parsons
```

This letter arrived when Joseph really needed some positive reinforcement. Work with the power company continued, but it was difficult to get supplies. An embargo was ongoing on shipping due to the war. Dwight Hopkins owned the lumberyard in town. He was a self-made man of some means, and he knew how to cut lumber. One day, he decided he should generate his own electricity. He purchased a small motor and a big boiler and called Joseph to wire it for him.

Seeing the way he had hooked them up, Joseph said, "This is never going to work in your sawmill. It's much too small a motor."

"Oh, sure it will!" Mr. Hopkins replied. "We'll just fire up the steam and feed the wood to it. It will run!"

Joseph decided not to argue and wired the motor to the saw. While one of Mr. Hopkins' men filled water into the boiler, another filled the burner section with scrap wood. Steam pressure quickly came to pressure, and the saw was running at a good speed. Finally, the switch was thrown on the conveyer feeding the first log down to hit the blade.

Bang! There was an explosion, and that was the end of the generator.

The sawmill still needed its own power, so Joseph agreed to build a plant that would work.[97] As part of that power plant, a quantity of copper wire was ordered from the Safety Insulated Wire and Cable Company in New York City. Unfortunately, they could not ship it because the railroad had been allocated for shipment of war materials only.

---

[97] Joseph Franz, "Direct Connection of a Motor to Swing Saw Permits Use of Smaller Size," *Electrical World*, 1916.

Not to be stopped, Mr. Hopkins declared, "We'll just go down to New York and get it." This led to an amusing adventure for them as they traveled to New York City.

Mr. Hopkins owned a Stanley Steamer automobile, so they were going in his car. It would occasionally run out of water, which meant they had to stop by a stream to fill it up again. Unfortunately, while they were on this trip, there wasn't anything in the car to carry the water, except a hat. And that is what they used. They finally arrived at Poughkeepsie and decided to spend the night. The hotel they found was very elegant. It even had a parking attendant.

However, when Mr. Hopkins wanted to put his car in the garage, the attendant said, "You can't put that thing with an open flame in a closed garage!"

Thus, the car had to sit on the street for the night.

They registered for a room at the desk and then went into the dining room for dinner. After they were seated, a waiter brought a tray of appetizers. Joseph took a sampling of the food, but Hopkins took the tray and dumped the whole thing on his plate.

When they ordered the main course, the waiter asked Joseph, "Do you want something for him, too?"

It was embarrassing for both Joseph and Mr. Hopkins because Joseph was Mr. Hopkins' guest.

"I didn't know they were going to bring anything else, and I was hungry!" Mr. Hopkins later apologized.

Apparently, he had not eaten in many fancy restaurants. In those days, hotels did not have separate bathrooms with every room. There were washstands with a bowl and pitcher of water in the room, but the toilet and bath facilities were at the end of the hall. Joseph had the foresight to pack pajamas and a bathrobe, but Hopkins just slept in his long red underwear. During the night, he needed to use the toilet. He proceeded down the hall, clad only in his red long johns. Just at that moment, a lady was coming in the opposite direction. When she saw Hopkins, she let out a blood-curdling scream, thinking she had seen a devil. She ran screaming into her room. They did not stay overnight on the way back.

Several years later, Mr. Hopkins' son, Steve, took over the business. He was a bright young man and very respectful. Joseph's dealings with him were much easier and far more pleasant than the adventures with his father.

Joseph's reputation in the power industry grew during this period. General Electric asked Joseph to consult on a number of projects. For instance, he was chosen to try out their first static capacitors. In addition, he was given their new lightning arresters to test before large-scale manufacturing was undertaken.

Mr. de Heredia owned considerable stock in the Niagara Power Company. He had an idea that he could perhaps use a small stream on his estate to generate his own power. The Niagara people referred him to General Electric. Mr. Rosart, a sales engineer for General Electric, asked Joseph to evaluate the situation with him. In the resulting report, Joseph and Mr. Rosart explained that the water flow of the stream was not sufficient to generate power.

At another time, Mr. C. E. Chesney, manager of the Pittsfield General Electric plant, called Joseph to the Wendell Hotel and asked his advice about the City Savings Bank investing in the Telford plant in East Lee, which was to get its water supply from Goose Ponds on Kinsbury Mountain. Mr. Henry W. Taylor, a consulting engineer, was quoting three million kilowatt hours of output for the project per year. The promoter, Taylor, had previously tried to sell the scheme to C. D. Parker and Company. Joseph calculated, from the watershed of about forty-three square miles, that the total output generated would only be around one million kilowatt hours. This information was conveyed to Mr. Chesney. As a result, the Pittsfield Savings Bank did not go into this venture. However, the treasurer of the bank, Mr. Ford, put his private fortune into it. He also sold stocks and bonds for the project throughout Berkshire County. Mr. Taylor proceeded to build the East Lee power plant.[98]

Ironically, during the first year the plant operated, it really did turn out the three million kilowatt hours. Mr. Taylor cleared out with all of the proceeds he could get. It took more than two years to refill the ponds. Later, Joseph's figures proved to be correct. Stockholders and bondholders lost their shirts. It is sad how apparently intelligent people are so easily cheated. The Mountain Mill Paper Company took over the plant to fulfill a guarantee to deliver one million kilowatt hours to the mill for taking their water rights.

In 1919, the Parker Company was in the hands of the bankers, who had control of several other small companies in Massachusetts. They eventually had fifty-two companies. A crew of accountants did all of the bookkeeping.

---

[98] "Use of Small Water Powers Simplified by Induction-Generator Plant," *Electrical World*, March 25, 1922.

There was one chief engineer, George Perry, and an assistant, Henry Parsons, whose brother had charge of the Palmer Light and Power Company. The Boston office had to approve all expenditures for new construction. Bowen Tufts, vice president of all of these companies, managed the entire enterprise.

Monthly meetings were held in different places. During the summer, as is typical of big business, the meetings centered on playing golf. Joseph joined the Stockbridge Golf Club and learned to play. The entire day was taken up figuring the handicaps and having lunch and dinner. Prizes were then announced. Finally, when everyone was thoroughly groggy, business was discussed. It consisted mostly of pep talks about selling appliances, window trimming, and, occasionally, construction policies.

By the end of the war, the C. D. Parker Company conceived a merger of all of their holdings into one combine, the Massachusetts Utilities Holding Company. New stock was again issued. C. D. himself received his usual commission. As before, for each share of stock held, the stockholder got so many shares in the great new combine. Because all of these transactions were just on paper, the old individual companies still operated in their usual way.

Beside his work for the electrical company, Charley Hull was good at getting Joseph to do miscellaneous jobs around town on company time. For example, the Benjamin Drug Store building once stood right at the end of the marble stonewall of the Episcopal church. Many of the members of the church were opposed to having the commercial building so close to the church. Money had been raised to buy the building and tear it down in order to build a parsonage. Mr. Benjamin did not own the building where his store was located. Even though he bought a block across the street,[95] he felt his drug store should stay where it was. It was across from the Red Lion Inn. He derived considerable business from there. The town insiders went to work. The original proposition was to build a stone parsonage in order to match the church. The church had already acquired the drugstore building. After several meetings of the vestrymen, including Charley Hull, who was the treasurer and a deacon of the church, and Mr. A. T. Treadway, Mr. Benjamin, and Mr. Punderson, it was decided to move the commercial building to its present location, specifically, across from the town office building on Main Street. At the same time, two of the Seymour houses were moved onto the church land and remodeled. One would sit on the back of the property as a parish house. The other would sit where the block had been. It would be a house for the rector, even though Miss Virginia Butler had previously built a fine house for a rectory across the street.[99] Charley asked Joseph to make all the plans for the rearranging and alterations.

When Joseph finished the plans, Mr. Hull suggested that Mr. DeGersdorf, an architect and member of the congregation, should review the plans. While Mr. DeGersdorf did review the plans, he didn't make any changes, but he still sent a bill for $1,000 to the committee.

Miss Mabel Choate, daughter of Joseph H. Choate, told Charley, "Of course, Mr. DeGersdorf [her cousin] could not work for nothing."

Joseph got nothing. His only recourse was that it was done on the time of the Stockbridge Lighting Company. After all of the buildings were moved and settled, the Episcopal Church wanted to rearrange the church interior. They wanted an organ on the lower floor as well as space for the choir. The baptismal font and pulpit were on the wrong side. Joseph designed all of the plans for these changes. Again, much of the work was completed on company time, even though he never quit at five o'clock. When he started a project, he worked into the late hours of the night. He didn't stop until it was finished. Nevertheless, he never received any overtime pay as compensation.

Joseph later surveyed Charley's property and a wood lot that was sold to the Proctors. On that job, he lost his plumbob.[100] Charley most generously offered to buy Joseph another one.

During these years, both children were in school and had made many new friends in town. With more consulting jobs and less responsibilities with the power company, Joseph had more time to be with his family. He began enjoying individual sports, like skiing, ice-skating, and tobogganing. They also had enough money to purchase equipment needed to take part in these sports.

For Christmas in 1916, Santa Claus brought the children skis with special harnesses to hold them onto their boots. It was almost a disaster though. Santa had forgotten to bring the skis in from where they had been hidden. Joseph raced to bring them in before the children came downstairs.

---

[99] A. T. Treadway later purchased the Butler rector house for his son Heaton at a favorable price.

[100] A plumbob is a pear-shaped lead weight that ends in a point and is suspended from the end of a line. It measures water depth or determines if a wall is vertical.

In those days, the family lit its Christmas tree with candles, which was very pretty. However, they had to stand by with the candle extinguisher and a bucket of water to douse any potential fire. When strings of lights were available, they switched to electric bulbs.

The Heaton Hall golf course was built on a fairly steep hill across the street from the old hotel. In the winter, when there was enough snow, Joseph would build a toboggan run. Anyone who was sports-minded in town could use it as long as the snow lasted. After school, on weekends and vacations, Natalie, Russell, and Joseph would be there. Besides the toboggan run, the hill was good for skiing, particularly for youngsters and those with limited skiing experience. Russell would try anything on skis.

Russell had three great loves that lasted his entire life: skiing, golf, and miniature trains. For Christmas when he was four years old, Joseph bought an elaborate steam engine that worked on real steam. It was fired by an alcohol burner, and it had a working bell and whistle. It ran on the bare floor because it didn't have a track with it. The next year, Russell received a coal car to go with it. It was hard to tell who loved that train more, the father or son.

When he was around ten, Russell began working as a caddy at the Stockbridge Golf Club after school, on weekends, and during summer vacation. Besides earning money, he learned to play the game and eventually did quite well.

Russell once picked up a stray cat on his way home from school. For a while, it was his favorite toy. It slept with him every night until his parents discovered the cat was infested with fleas. Russell woke up with red welts all over him. When the kids left for school that day, Joseph got rid of the cat. It was sad for the children, but the poor cat was beyond help. Natalia had to boil Russell's bedclothes to get rid of the fleas.

Joseph still had his original racing skates. When the Stockbridge Bowl and other lakes and ponds in the area froze over, he took the children and shoveled off the snow so there would be an adequate space for skating. Other people often enjoyed the skating rinks as well.

Figure 22—Joseph skating with his racing skates

Snowshoe parties were sometimes organized with friends. The group would tramp over to Rockdale, where they would have hot soup and cocoa in the power plant and then return home. Other treks took them through nearby woods. After which, they would end up at someone's home for hot refreshments.

In the summer, there was hiking in the hills. They'd go swimming, mostly at the Stockbridge Bowl. They sometimes had leisurely picnics. Sometimes, they'd go to Lake Buell in Great Barrington or to the top of Mount Greylock.[101]

Charley Acley, the eventual proprietor of the Benjamin Pharmacy, was one of Joseph's good friends. He owned a small plot of land on the south shore of the Stockbridge Bowl, next to the town beach. Their families got together for picnics at the lake. Their children played together and learned to swim. Charley let Joseph and his family use the lake site whenever they wished because Joseph would mow the lawn, clear away the poison ivy, and help paint the changing rooms, outhouse, and main cabin.

The many hills and mountains in Western Massachusetts offered myriad trails for walking or cross-country skiing. Their summits offered spectacular views during the entire year. The fall was the most colorful. Unsurprisingly, so many artists and scientists were attracted to this part of the country. Nature alone offered so many possibilities that the war in Europe still seemed very far away.

---

[101] Mount Greylock is the highest Peak in Massachusetts.

Chapter Thirteen

Power and the Roaring Twenties

When the armistice was signed on November 11, 1919, the war in Europe was officially over. President Woodrow Wilson's efforts to form the League of Nations came to fruition in Paris, but he was regrettably unable to convince the United States Congress to join it. Americans were still too isolationist.

Except for a small army of occupation, the soldiers returned home. John Foley from Stockbridge was one of the troops who stayed in Europe. He had been part of Joseph's crew before the war. From the letters he wrote to Joseph, describing places he had seen in France in great detail, he was obviously taking advantage of this opportunity to travel in Europe. He was most impressed with Versailles when he was there, just before the peace treaty was signed in 1919.

News from Joseph's family in Europe finally started arriving again after the war. Food was in very short supply, and living conditions were tragically meager. The family in America sent food packages and clothing for several years. Alma, Joseph's youngest sibling, wanted to come to New York City and asked for money to pay her way over. Karl's daughter, Teddy, tried getting the rest of the family in America to contribute to help Alma, but no one had enough resources to spare. Alma never came.

Ludwig had been recalled into service, and the Russians captured him very early. For three years, the family never heard from him. In those first years, he was forced to help build factories in Kazakhstan without any pay. When the war ended in 1919, he could write and let his father know he was alive and working in a factory. Because he was a chemist, the Russians had him experimenting with new ways to use fats. He was allowed to return home to Vienna in 1925.

In the United States, Warren G. Harding became president. Prohibition began, and women received the right to vote. As soon as governmental restrictions of goods and supplies were lifted, there was a return to materialism and the acquisition of wealth. Like other companies, the power industry looked for new ways to make money.

This ushered in the era of the banker's ownership of the Berkshire District. More stock would be sold. It was the duty of every loyal employee to sell all of the stock he could to his relatives and friends. In 1920, the Berkshire District became the Southern Berkshire Power and Electric Company. Charley Hull continued as the president.

Joseph's working conditions remained the same, even though changes occurred in the company structure. Complications in the power picture arose when a group of bankers known as Harriman Trust[102] sought to eliminate the small Western Massachusetts electric companies that were surrounded by their holdings and consolidate them into their company. Their plan was to obtain fifty-one percent of the stock. Then, the rest of the stockholders could be forced out. At the time, all stock was sold at a definite price, far above the market value of the individual stock.

C. D. Parker devised a scheme to form a voting trust. All of the stockholders were to sign over their right to vote for a considerable monetary consideration to a set of voting trustees. The right to vote would be restored if the stock failed to pay the regular dividend. The voting trustees were all of C. D. Parker's men. The arrangement seemed fine and noble. The company had many years of meager income, but the dividend was at least five percent. As a result, the consolidation occurred quickly.

The resulting New England Power Company was a "wheel within a wheel." The New England Engineering and Service Company[103] was the parent company. It was not incorporated. It was simply a partnership. Such organizations were not subject to any corporate laws or commissions. It was a consulting engineering and contracting firm that did all of the engineering work for the various underlying utility companies. A purported advantage was that it saved the individual companies a large amount of money. One central administrative staff could supposedly more efficiently handle engineers, construction personnel, manual labor, and equipment with less overhead costs.

Despite abuses that crept into the large organizations, there were some good qualities to offset them. The original small electric companies were similar to the Stockbridge Water Company[104] that remained independent. They were

---

[102] New England Power Company

[103] Joseph was a director and also an engineer.

[104] B. I. Moore, "It All Happened Here," *New England Electric System*, central edition (1957).

ultraconservative. Only concepts that were tried out very well were accepted or carried out. Then, they were only used in very narrow, limited channels. The men who were at the head of these small corporations were usually picked because they were successful in some other business. It was taken for granted that these persons would be suited for the utilities field as well.

That sort of management did not permit progress in a highly scientific or technical enterprises. Bankers were still the moving spirit in large corporations. During their reign with the C. D. Parker Company, they were at least willing to spend and loan money for new developments and expansions.

Young Charles Bidwell from Stockbridge was an MIT graduate. Joseph broke him into the electrical business during his summer vacations. He once read meters for Joseph. Charley Hull picked him for future manager of the Great Barrington Electric Company. Mr. Tufts, head of the power company, once said he was unassuming. That seemed to be the major requirement for a man in public utilities development. Bankers wanted men who would be subservient to their will. Carl, as he was known, surely fit that requirement. Whenever a question arose about something he did not know or could not resolve, he would take it a higher-up. As a result, he never got into trouble.

One of the follies of such an arrangement was not giving the subordinates the power of discretion. One example is an incident in South Egremont. The distribution lines were run on private property off the main highway. One pole was in a Mr. Crissey's chicken yard, and it became necessary to do some work on this pole. Mr. Crissey, another shrewd Yankee, claimed the men had left the chicken yard door open and his prime rooster got out. A passing car on the highway then killed the bird. He asked for five dollars for the loss of his rooster. Carl Bidwell called his boss, Mr. Dustin, and explained the problem.

"Two dollars is all we pay for a rooster," Mr. Dustin decided.

Therefore, Mr. Crissey was given two dollars. Mr. Crissey made a good living because he owned a feed mill and sold grain. He knew exactly how he could collect. He filed a lawsuit, and it wasn't for five dollars. The bird had now become a very valuable prize rooster worth twenty-five dollars. Summonses were served. Carl called Mr. Dustin again. Mr. Dustin called Boston to send their corporate lawyer to defend this suit.

The Boston answer was, "Pay it! No one from here is going to come to Great Barrington for any twenty-five dollar lawsuit."

The suit was paid, plus court costs and telephone bills.

In the Lenox Branch, T. J. Newton was absolutely unfit for his job as manager, and he was finally fired. In the original sellout, T. J. had been promised that his son, Elmer, would have this soft berth. However, much to Joseph's pleasure, that didn't happen. Instead, Mr. Alexander Wylie Jr., an accountant, took over in Lenox.

Financially, Joseph was at the height of his career. He received a bonus for all power generated by their plants, and he had a salary as well. He made as much as $6,000 per year in salary, and his private business netted from $10,000–15,000 per year. He was promoted to power engineer of the Berkshire District.

In 1922, part of Joseph's job was, as coengineer with Charles T. Main, to be in charge of construction and installation of the fifty-kilowatt Rockdale Powerhouse on the Williams River in Williamsville, Massachusetts. It was the first successful, entirely automatic, hydroelectric plant.

Figure 23—First automatic hydroelectric plant on Williams River in Williamsville, Massachusetts

The first automatic control for a powerhouse, installed by General Electric, had failed. It consisted of a rubber ball in the forebay. A copper tube from this ball went to the regulator at the plant, some 600 feet down on top of the penstock. Due to temperature and pressure, expansion made this system impractical. At the Rockdale plant, Joseph installed a float switch in the vent pipe, filled it with kerosene, and added a rheostat. His system worked perfectly.

By then, Joseph could boast he sold more horsepower in General Electric motors in one year than any other General Electric motor salesmen. General Electric printed stationery for Joseph, along with blotters he could send with his letters. All other privileges of doing work on his own remained. He was selling and installing household appliances, such as large electric ranges and walk-in refrigerators in the larger houses and dairies. This caused some friction with the Lenox manager, Alex Wylie, who was also a General Electric's sales manager. While Charley Hull was still president and manager of the Stockbridge Division, regardless of how much Joseph did to prove his worth, he was never promoted to a higher administrative position in the power industry. Joseph could never become more than the power engineer.

Joseph met many wealthy people when he was selling electric appliances and equipment. He visited his friend, Charles S. Mellon, often at Council Grove in Stockbridge and at his home in Concord, New Hampshire. Joseph enjoyed his company because they could talk about many different subjects as equals.

Mr. Mellon often told Joseph, "I was only an office boy of J. P. Morgan."

They were once instrumental in starting the incorporation of a savings bank in Stockbridge. It was dissolved when the Housatonic National Bank agreed to put in a savings department in the Stockbridge branch.

Joseph did a large amount of work for Mr. Mellon. During the war years, they built a complete grist and flour mill along with a modern dairy and cheese-making plant on his New Hampshire estate. Electric pumps, water supplies, electric stoves, refrigerators, and other appliances were furnished and installed.

Joseph's first meeting with W. W. Field, husband to the former Miss Lila (Vanderbilt) Sloan, was when electric service was being supplied to their new residence at Highlawn Farm. During the years after that, Joseph sold him all sorts of electric equipment, including one of the first radio sets with a 150-foot steel antenna pole that sat right out on the front lawn.

Joseph called on him often and had the liberty of the house, providing he came in through the butler's pantry. Joseph quickly learned he was not considered a social equal. The same protocol applied in other cottages as well. If he came to the front door, Mr. Field was never home. Joseph could stand this humility because Mr. Field spent his or his wife's money generously. If Joseph came in the morning, he'd often meet Mr. Field at breakfast and have coffee, honey, and rolls with him. Joseph enjoyed talking with him. Though a graduate mechanical engineer, he knew little of practical engineering. However, Mr. Field had many hobbies, including dairy farming, bookbinding, photography, woodworking, agriculture, and horticulture.

Joseph installed a twelve-plate Du Parkay kitchen range with four ovens,[105] an electric walk-in refrigerator, pantry-warming ovens, and a complete dairy with a specially constructed milk carrier to transport the milk from the stables to the dairy. Additionally, he designed an electric hay hoist. Beside all of these things, Joseph electrified their water-pumping plants. The water supply came from a spring under the mountain more than a mile from the house and barn. The storage reservoir was on top of the mountain. The new electric pumps were Gould's with automatic controls that replaced the former steam pumps. All of this was private work, including the transmission lines for the electric power supply.

In Mr. Field's den, Joseph once asked if his mother-in-law, Mrs. Sloan, the former Emily Thorn Vanderbilt and widow of W. D. Sloan, would be interested in an electric kitchen such as theirs.

Mr. Field called, "Lila, do you think your mother would be interested in an electric kitchen range?"

Mrs. Field responded, "Well, Billy, if you speak to mother, she may consider it."

He sarcastically replied, "No, thank you! I'm not going to compete with that butler of hers."

Mr. and Mrs. Fields sometimes came up during the off-seasons. They stayed in the so-called playhouse. For a day or so, they kept house by themselves. Mr. Field often made flapjacks for breakfast. Joseph would turn up soon after they arrived.

---

[105] The cost was $2,500.

"This is uncanny! How did you know I was here?"

Mr. Field never did understand how. Joseph never let him know where he got the information. It was simple enough. The superintendent always knew when his boss would be there. Mr. Field simply thought Joseph had a sixth sense. Usually, there was another order for something.

On one occasion, Mrs. Field took Joseph aside and said, "Please don't sell my husband anymore things."

Joseph wasn't always treated like a tradesman or one of the help. As a member of the AIEE, Joseph was treated as an equal by such eminent men as Giuseppe Faccioli, chief electrical engineer for General Electric in Pittsfield, and the world-renowned electrical wizard, Dr. Charles P. Steinmetz. As members, they often went on scientific field trips that experts in a particular field led.[106]

During this time, life for the Franz family was generally good. By 1922, the Chrysler was just not reliable enough anymore, so Joseph bought a Franklin for family use. That was a grand vehicle! It had a plush interior with jump seats in the rear as well as a footrest for the backseat. His business of selling General Electric motors, farm equipment, and large household appliances had greatly increased. The same year, he bought himself another car, a Model T Ford coupe to provide more mobility.

Though he never learned to play an instrument, not even his banjo, Joseph always enjoyed music, particularly opera and classical music. Happily, the Colonial Theatre[107] in Pittsfield regularly presented touring companies. Joseph took the family to hear Marion Talley, Madame Amelita Galli-Curci, Louise Holm, and others. The children sometimes got bored, but, if nothing else, they absorbed something by osmosis. Joseph bought a manual Victrola and a record collection so they could listen to the great singers and musicians at home. They even had a few recordings of music hall comedians, like Harry Laughter, which were always good for a laugh.

Unfortunately, Natalia's health continued to deteriorate. She began having more gallbladder attacks, causing a great amount of pain. She lived on one kind of medicine or another.

---

[106] "The Eclipse," *New York Times*, Sunday, Jan 25, 1925.

[107] This landmark is currently being restored.

Yet many good times were enjoyed as well. On weekends, they went for a drive somewhere or visited friends.

Figure 24—Joseph and Natalia Franz, 1920

In addition, if they were at home, their house seemed to be always filled with people eating, drinking, singing,[108] and laughing. Joseph sometimes wanted a little peace and quiet alone with just his own family.

---

[108] Natalia loved to sing.

The children, particularly Natalie, occasionally went with their father when he consulted clients or visited professional colleagues. Natalie even attended some of the AIEE meetings when she was in high school. Russell was too busy with his friends to want to tag along with his dad very often. Everyone in town knew and liked Russell. He was easy to like, but he was lazy in school and often just barely passed. Joseph scolded him when he brought home his report card with bad grades. The grades would be higher the next month, but his parents believed the teacher just loved Russell's happy smile.

Natalie was a good friend of the Mellon's daughter, Priscilla. In the summer, they swam in the Mellon's pool and rode horses together. In the winter, they went ice-skating on their pond. The Hull's were apparently somewhat jealous of this friendship because their daughter, Charlotte, was never invited.

One day, Natalie reported to her parents that Mrs. Hull told her, "I don't want you to go up to the Mellon's any more."

It was a terrible thing for her to say to someone else's child, but it did not stop Natalie from seeing her friend. Years later, after her father died, Joseph would sometimes visit Priscilla when he was in New York City. While he enjoyed his visits with her, he could not talk to her like he had with her father.

On November 24, 1923, a great personal tragedy struck the family. Natalia was stricken with a very severe gallbladder attack. He frantically called Dr. Dodds, who came as quickly as he could.

When he saw her condition, he declared, "She'll never make it to the hospital.[109] I'll have to operate right now!"

He ordered Joseph to wash the kitchen table and prepare boiling water to sterilize his instruments. They carried Natalia downstairs and laid her on the table. He had some ether, but he didn't have a mask. Joseph retrieved a strainer and a clean dish towel to anesthetize her. The doctor scrubbed his hands and told Joseph to do the same in order to assist him. Joseph boiled the tools in another towel, so he could take them out of the pot without touching them.

Then Dr. Dodds went to work. Joseph's head spun as he saw the doctor cutting into the person he loved. Miraculously, he managed to get through the ordeal without passing out himself. Blood was everywhere. Suddenly, she stopped breathing. The sweat ran off his forehead as Dr. Dodds swore.

---

[109] The hospital was in Pittsfield.

"Damn!"

Joseph knew she was dead. The gallbladder had ruptured. The doctor couldn't do anything to save her under such primitive conditions. It was doubtful she would have survived, even in a hospital. Joseph cleaned up the kitchen before leaving so the children would not see the gruesome scene. They took her body to the hospital for the death certificate, as the law required.

Russell and Natalie were home from school when their father returned from Pittsfield. They were fifteen and sixteen years old. They were old enough to understand the sadness, but they weren't old enough to witness evidence of the horror of their mother's death. Joseph told them that Natalia had died on the way to the hospital. It was a terrible blow to all of them. Of course, life would go on, but it would never be quite the same again.

Chapter Fourteen

The Healing

The next couple years were very lonely and depressing for Joseph, but there wasn't any time for personal commiseration. He was swamped with work, and his children needed a mother. He tried spending as much time as possible at home. Natalie started cooking. Russell was supposed to help with various chores, but he managed to evade most of them. Joseph could not share his sorrow with Natalie and Russell because they had to deal with their own grief.

Disturbing news from Europe compounded the trauma of Natalia's death. The war had been devastating everywhere. In the aftermath, there were still food shortages, lack of essential medical supplies, and poor transportation. Jobs were hard to find. The economy was inflated. The things that were available were very expensive. Joseph's father wrote, informing him that they were mostly living on potatoes. They were only allowed 250 grams[110] of meat and a quarter-pound of fat per week for the whole family. Family members in the United States continued sending what they could, but it was never enough.

Within the year after Natalia died, Joseph began looking for someone to provide the emotional relationship that both he and the children needed. In a letter, Joseph's father said he knew how his son felt because he had felt the same way when Joseph's mother died. He hoped Joseph would find someone to share his life again. Attempting to help his son, he enclosed a picture of a Lina Müllner, the sister of his boss in Vienna. He described her as being single and a professional bookkeeper who knew English. She had a strong, honorable personality. She was willing to come to America and might be just right for Joseph. He also wrote that Aunt Elise was visiting from Heidleburg, and Uncle Josef was planning to come to Vienna that summer. He hoped Joseph would be able to come home at least once more. Joseph appreciated his father's concern. Now he understood why his father had remarried. Nonetheless, he was not interested in a long-distance romance, even with an attractive lady.

---

[110] 250 grams is one-half of a pound.

Joseph was over forty, but he was still strong and virile. Finding the right single woman was not going to be easy. Some of the schoolteachers, like Gertrude Wolfe, were relatively pleasant company in small doses, though, in general, it seemed the most intelligent, good-looking women were already married. The widows were too old, and the old maids were either physically unattractive or did not have the kind of personalities Joseph wanted.

One German family in town, named Radell, had two unmarried daughters. Joseph occasionally saw the girls tobogganing or skiing on the Heaton Hill golf course. He had once even ridden on the same toboggan as Emilia, the older sister when they were in their late teen years. But Emilia didn't pay any attention to him. He was just another married man to her. On her father's side, she was a second-generation American. She was a first-generation American on her mother's side. She had an appealing air of refinement and self-assurance. Joseph and Natalia had been active in the Community Service Club[111] that met at the old Casino on Sargent Street. The Radell girls, like many other Stockbridge residents, also participated. Emilia took parts in many of the variety shows and pageants that were staged every year.

One time, the club arranged a walk to Interlaken,[112] about three miles from the center of town. Emilia passed Joseph on the way back. It was relatively late, so he offered to see her home.

"No, thanks!" she replied. "I can walk home quite well by myself."

She seemed like a bright girl and was very independent. She was not glamorous, but she had a kind of inner beauty that originates from self-confidence. She knew who she was and where she was going. She had studied art in Boston and Chicago and had taught the subject at schools in Lanesboro and Stockbridge.

When Joseph tore down the old Miller house on Elm Street, Walter Paterson, the local carpenter, bought a couple of the doors. Walter then installed them into a small house on Church Street that John Radell, Emilia's father, purchased for his family after he had retired from his position as care-

---

[111] Joseph served a term as chairman of recreation. During which time, he arranged for a performance of the Chautauqua Circuit to perform in Stockbridge. He later served as chairman of winter sports and built the town's ice skating rink near Park and South Streets.

[112] Interlaken is another district in Stockbridge. It is where the first pulp paper mill in America had been. See article in *New York Times*, January 25, 1915.

taker of the Southmayd[113] estate. This was the first unknown connection between Emilia and Joseph.

Joseph had not seen Emilia around town for some time. She had been working in Washington, DC, during the war. Then she had taken an extended trip to Europe with her sister, Frieda. Early in 1924, Joseph spotted Emilia, then thirty-five years old, walking along Main Street. He decided to get to know her better. He found out she was working at the Riggs Foundation, teaching pottery and weaving in their shop across Main Street from the Brahman block. Once, when she was walking home to Church Street after work, Joseph was driving down Main Street. She looked tired, so Joseph offered to give her a ride home. She gratefully accepted.

During the first week in April, a snowstorm blanketed everything. Joseph decided to go for a sleigh ride, but he really did not want to go alone. For the first time, he called Emilia on the phone. To his surprise and delight, she said she would love to go for a sleigh ride. She had the same idea when she saw the snow. On the way, he let her drive the sleigh, mainly so he could put his arm around her. He probably was too hasty, but it had been a long time since he had been alone with a woman. She was not very receptive. The snow melted quickly, and the runners of the sleigh were already scratching on bare spots by the time they returned.

He called a few days later to see if she wanted to go out for a drive, but she refused. In the next three weeks, Joseph kept calling. He'd also see her on the street. She always refused to go with him. One day, he saw her walking on East Main Street when he was headed in the same direction by car. He stopped to see if she wanted a ride. She said she couldn't because she was on her way to Lee.

"Well, get in," he offered. "I'll be glad to take you take you over there."

She let him drive her to Lee, but she insisted she would get back on her own. When she had a bad cold that spring, he sent her flowers and candy. She sent him a very nice thank-you note.

By the end of summer, they were seeing each other more frequently. During the rest of that year, they went to dinner at the Great Barrington Inn, the Oaklawn Inn, and other places in the area. They took rides to view the fall leaves and walked on the mountain trails. When Joseph was out of town on

---

[113] Charles F. Southmayd was a partner of Charles E. Butler, Joseph H. Choate, and Evarts.

business, he wrote to her and sent her candy or flowers when he returned. The snow arrived early, so they went skiing, tobogganing, or sleigh riding. They also took little trips to view the winter landscape.

In December 1924, Emilia's eighty-four-year-old father was seriously ill. Emilia was also plagued with constant colds and coughs. The doctors said it was because of infected tonsils. Joseph had proposed marriage, but Emilia did not want to commit to an engagement until she cleared up her medical problem. Coincidentally, her Uncle Louis Ludwig in Washington, DC, was involved in a terrible auto accident, which landed him in the hospital. Emilia volunteered to care for him when he came home. She could get her tonsils out in Washington, DC, when he was better. She also wanted to talk to him about her possible marriage.

While she was away, Joseph wrote letters to Emilia most every day. At first, he generally reported news of her father and other family members, along with local events, including occasional entertainments, births, and deaths. That winter, there was an unusual number of deaths of both relatives and friends. Emilia had been on her own for many years and had many friends in Washington, DC. Her infrequent letters told of seeing many friends, the parties given for her, and dates to the theatre and concerts. In frustration, Joseph wrote of his concern that, because she was enjoying other men's company so much, she must surely be thinking of eloping with one of them. He begged her to write her true feelings for him, to reassure him they still had a chance for happiness together. After a month-and-a-half, his letters suggested that enough time had gone by to attend to her uncle and her mother really needed help taking care of her father.

Fortunately, Emilia's sister, Frieda, could spend a week or more at home to relieve her mother. The visiting nurse was coming regularly. By mid-January, at Joseph's insistence, there was a family conference to see if Emilia should have the tonsillectomy in Washington or come home to have it done. The consensus was that she should stay in Washington because better doctors were there. On January 19, Emilia went into the hospital. The next day, Uncle Louis telephoned Stockbridge to let everyone know she was doing just fine. Joseph sent fifty dollars to help pay for any expenses[114] and told her he would be down to visit her after attending an AIEE field trip to see the eclipse of the sun in New Milford, Connecticut.

---

[114] Emilia did not want to accept his money.

After viewing the eclipse on January 24, 1925, Joseph took the train to Washington to see Emilia in the hospital. She enjoyed hearing about Joseph's experience. Arthur Palme was the tour leader and one of those who photographed the phenomenon. Between fifty and seventy-five men and women gathered on a particular hillside for the event. People were warned not to look directly at the sun because it might damage their eyes, even during the full eclipse. Some people had dark glasses and took a chance anyway. Most did as they were told and made a small hole in a piece of paper. They turned with their backs to the sun and held the paper in front of them. The shadow of the sun was projected onto the white snow. It made a perfect image of what was happening in the sky.

Joseph stayed in Washington until January 27. He was very grateful for the many kindnesses and considerations Louis had showered on him and for all he was doing for Emilia. He hoped to reciprocate when they were settled.

Joseph's letters grew more and more emotional.

"The suspense is terrible. However, don't blame me too much for getting impatient. I hope you do not think I am not interested enough to see you all well. Nevertheless, remember, all you ever wrote me was 'Now I am well and do nothing but eat and sleep.' Then you wrote me about going here and there. Naturally, I am a little impatient. I think if you can do all these things, considering the long time since your operation, you surely were neglecting me just a little. At least that's how I feel. Now, if I did not care for you, as I do, would it matter where and how long you stayed?"

On March 6, Emilia finally returned home. During the time in Washington, not only did she recuperate from her operation, she bought her wedding dress and trousseau for her coming marriage.

In the spring of 1925, Joseph had several private surveying jobs. One was making the cemetery map that hung in the town offices. Emilia came along and helped him with the survey. She also did all the lettering on the map. Another job was for the Stockbridge Golf Club. Established in 1895, it was one of the first 100 in America. When more land was acquired, it was remodeled from a nine-hole course to an eighteen-hole course by Walter Nettelton. In 1924, another parcel of land was added, and Joseph was hired to remodel the course once more.[115] He produced a large, colored map of the finished course, and Emilia again did all the lettering. For many years, it was printed on the scorecards in reduced form.

---

[115] http://thegolfcourses.net/MA/1422.htm

During their work on the cemetery project, they got to know each other fairly well. She understood when Joseph talked in technical terms and sympathized with some of his work problems. She had also met his children. On one outing to Gilder Pond on Mount Everett, Joseph begged Emilia to please announce their engagement. She agreed, and the announcement was made on April 4, shortly before her father's death.

When her father died, Emilia and her mother agreed there wasn't a need for a prolonged engagement. She asked her Uncle Louis to give her away, and he gladly accepted. With a fair degree of formality, they were married in the Congregational Church on June 13, 1925.

Figure 25—Wedding of Joseph and Emilia Radell.
John Palmer, Alice Schilling, Ruth Berridge, Charley Hull, Joseph, Emilia, Frieda Radell, Charlotte Wilcox, Jane Peters, Ned Wilcox. Youngsters: John and Walter Sisson (Emilia's cousins)

Emilia's sister, Frieda, was maid of honor. Charley Hull was the best man. There were four bridesmaids and two ushers. Emilia's favorite cousins were the two ring bearers.

Their honeymoon was a summer in Europe. Going over, they took the *Ryndam* from New York City to Rotterdam. One night during the trip, there was a bad storm. It was so bad that the main dining room was awash. The next day, many people were seasick. Emilia and Joseph slept right through the storm and felt fine in the morning. From Rotterdam, they toured some of the interesting places in The Netherlands. They visited Volendam and Marken, where people were still in quaint costumes and wooden shoes. They took a canal ride on Amsterdam's many lovely canals and were most impressed with the Hague and its Peace Palace.

KLM was already flying passengers in Europe. Emilia was as eager as Joseph to try this new means of transportation. Before they boarded, the staff asked Joseph for their address and where they were staying locally.

They asked, "Who is your next of kin?"

Joseph was not so sure that the flight was such a good idea, but he did not tell Emilia until they had safely landed. When he did, they had a good laugh.

The flight from Amsterdam to Brussels was on a dual prop bi-plane.

Figure 26—KLM airplane from Amsterdam to Brussels

There were wicker seats with a spittoon between each pair, just in case. During the flight, a mechanic got out on one wing to tighten a wire brace. By today's standards, they were not flying too high off the ground. It was a perfectly clear day. The view was breathtaking. Miniature people were farming or on their tiny wagons and carts. Small bicyclists were heading down roads that lead to proportionally small villages with thatch-roof houses. In less than an hour, they were on the ground again in Belgium.

During the rest of the way, the train was their mode of transportation. Their next stop was in Nastätten, Germany, to visit some of Emilia's relatives who were millers, bakers, innkeepers, and farmers. The old Funkenmülle, once owned by Emilia's grandfather, was still operating in that Taunus Mountain town. Three kilometers away was the town of Mielhen and the Gasthaus zum Nassauer Hof, which cousins of Emilia still owned. It has been in the same family for several hundred years. In nearby Laufenselden, they met more of her relatives. Communication wasn't a problem because Emilia had learned German from her mother and it was Joseph's first language.

They spent two nights in Frankfurt before boarding another train south. En route, they stopped for a night in Brucksal to see where Joseph had been born. His Uncle Joseph and family warmly welcomed them. They still lived in his grandfather's house and worked in the store Joseph remembered so well.

Figure 27—Joseph with his Uncle Joseph Franz and other relatives

They also toured the lovely pink palace. After another day on the train, they arrived in Vienna. During this visit, the couple were housed with Joseph's father in Gumpoldskirchen, a suburban wine-growing town near the city. Karl Anton's second wife, Adele, was an attractive young woman. Joseph's two half-brothers, Otto and Karl, were active five- and seven-year-olds. Emilia helped make their visit amiable so Joseph and his father had a warm reunion.

It was an easy commute by trolley to the city of Vienna, where they visited Joseph's brother, Ludwig, and his wife, Pepi. The palace of Schoenbrun was still as beautiful as Joseph remembered. Even though the city had changed, they managed to find and explore many other sights that Joseph had known as a boy. They gorged themselves on delicious Viennese cuisine, such as famous pastries slathered *mit schlag*[116] and imbibed in the *heurigen*[117] wine that was made in the region. One could always tell who was selling the wine because pine branches would be hung outside the door where wine and food was available.

Joseph's sister, Rosa, entertained them in the small house she had built in Gumpoldskirchen, not far from their father's place. The youngest sister, Alma, married Theodor Fritz in 1920, and they lived in the city. They had one daughter, Rosa. She was always referred to as Roselein or Roserl (Little Rosa) in deference to her Aunt Rosa, for whom she was named. When Joseph and Emilia visited, Alma seemed terribly frail. She died the following April.

Even though it was the last time that Joseph saw any of the family, they continued to write whenever they could. Joseph's father had been granted his wish to see his son once more before he died in January 1926. Emilia and Joseph were happy to see that life in both Germany and Austria seemed to have returned to more pleasant conditions, at least for their relatives.

From Vienna on another train, they crossed the Alps to Italy. Venice was their next destination. As typical tourists, they rode in gondolas and tramped the narrow streets to see the Piazza San Marco and its cathedral, the Doge's Palace, the Bridge of Sighs, and several museums. They bought some of the famous mosaic jewelry, glass trinkets, and lace. The northern Italian cuisine was a delicious change of menu from that north of the Alps.

---

[116] whipped cream

[117] new wine

The last leg of the trip took them through Milan to Bern and Lucerne in Switzerland. The stops were only long enough to view the spectacular peaks and quaint countryside. Joseph managed to buy a new coat in Lucerne, but he left it on the train to Paris, their final stop. Needless to say, he was very upset with himself. He had never even worn the coat!

While in France, they took a tour to Fontainbleu for a look at how the French built their palaces. This was a special treat for Emilia because she had not seen it on her first trip to Paris. Another must was a meal of the famous escargots,[118] a first for Joseph.

He had to admit, "They really aren't so bad!"

Additionally, Emilia had to do a little more shopping!

For the voyage home, they boarded their ship, the RMS *Berringaria*, in Cherbourg. The journey had taken more than two months to complete. They had shared a unique adventure. So many happy sights and sounds would take months to sort out, but the memories would certainly last forever. As history played out, the uncertainties of the next two decades made such travel hazardous or completely out of the question.

---

[118] snails

Chapter Fifteen

Power, Politics, and More

Joseph's first political participation had been with the Stockbridge Chamber of Commerce, which he joined in the early 1920s. He was president for two years in 1925 and 1926. At the time, there were sixty members. Allen T. Treadway, the Republican representative to Congress from the Berkshires, would occasionally come to their meetings when he had a bill he favored to be passed. He'd ask the Stockbridge Chamber of Commerce to pass a resolution in its favor. It would then be forwarded to various congressmen and senators. During his long career in the House of Representatives, he was a shrewd politician, even though he only initiated one bill that passed. Still, he was considered a fine representative because he'd go to no end of trouble to get a soldier's widow her pension. If one visited Washington, his secretary would provide passes to all the different government buildings. Sometimes, his chauffeur would even take one around the city. His staff sent first day of issue stamped envelopes to anyone who asked. He really enjoyed playing politics over the entire year.

Charley Hull was awestruck when Allen T. Treadway was around. Joseph could never figure out why this man—and later his son—held such power over the people of Stockbridge. Jim Punderson, Edgar Searing, and many others worshipped the ground Allen T. Treadway walked on. For example, at one town meeting when the question came up to build a new school, Jim Punderson stood up in opposition. The congressman was moderator. He shook his head at Punderson, who did not finish his sentence. Instead, he choked and just sat down. Several other incidents were like this, which puzzled Joseph.

During Joseph's tenure as president, one negative matter involved a proposition to appoint a town manager. The issue was brought up at a general town meeting. The reason stemmed from an unfortunate misappropriation of local tax funds. That problem had been resolved through the town's insurance company, but some of the chamber members felt a town manager might prevent such an incident in the future. The chamber formed an independent committee to investigate how town managers worked in other places. Committee members included Miss Mabel Choate, Dr. Lunt, C. Edgar Searing, Peter Adams, and Joseph. The committee went to several towns with town managers, including Waterviliet, New York, to talk with the selectmen and other town officials. Judge Hibbard was engaged to draw up proper bylaws based on what had been learned. For which, he charged $500. A plan to implement the appointment of a town manager was written, and it was presented at the next chamber meeting. The chamber members unanimously voted to recommend this plan to a special town meeting. The chamber's proposal was presented. Aside from those on the committee, only a couple other chamber members were at the meeting. Peter Adams had resigned from the committee and headed the opposition. Too many people objected to the person they believed would be appointed as the town manager. The proposal was defeated. Thirty were for it, and thirty-four were against it.

The next day, Joseph queried Charley Hull, "Why weren't you at the meeting?"

Charley responded, "I thought you could get along without me."

Obviously, many others did not go either.

Beyond rejecting the proposal, the town rejected paying the cost of the research. It cost each member of the committee $100 to pay the lawyer's fee, along with his or her personal expenditures for the research work. After that, Joseph dissolved the Stockbridge Chamber of Commerce. In recent years, another chamber has been formed, but there has never again been any mention of a town manager.

William Cullen Bryant[119] had lived in Cummington, Massachusetts, which is not far from Stockbridge. When they dedicated his homestead as a national monument in 1928, President Calvin Coolidge was the main speaker. Joseph rode over with Penn Cresson[120] and several others from Stockbridge. Penn was quite a talker and really wanted to talk to President Coolidge, but he could not get near him. Joseph never stood on much ceremony. He went right to the president and introduced himself.

---

[119] American journalist and poet (1794–1878)

[120] He married Margaret French, the daughter of the sculptor, Daniel Chester French.

"I remember when you were in Stockbridge having dinner at the Red Lion Inn. You were to speak in Great Barrington, and everyone drove off and left you sitting on the porch."

Cal laughed and started talking about all sorts of things.

On the way home, Penn told the others, "Here was Joe talking and laughing with Coolidge. All I got was a grunt!"

Politics was still tricky for Joseph in his career. The C. D. Parker Company was still in control of the Southern Berkshire Power Company. In 1927, Joseph designed and built an outdoor hydroelectric generating plant for them. George Perry was chief engineer at the Boston office.

He asked Joseph, "Did you ever build such a plant before?"

"Yes, I had something to do with such a plant in Switzerland," replied Joseph.

It was a white lie, but he received the authorization to spend $2,500 on the experiment. It was not a large amount of capital, and it was not a very large plant.[121] However, it was the first plant of its kind that was known to be successfully constructed anywhere in the world.[122] Built on the Williams River in the village of West Stockbridge, it was developed to make a marginal site financially attractive. When he was home from college for the summer, Russell Franz was part of the construction crew that built it.

---

[121] 30 kilowatt/2,400 volt/three-phase transformer

[122] It was in operation until 1948 when it was sold to the town of West Stockbridge.

Figure 28—First low waterhead outdoor electric generator

It was very simple to operate. A clerk in the general store across the street turned it on in the morning and off in the evening. Joseph's success was published in technical journals throughout the world.[123] The government of Japan even sent a commission to investigate the new power plant. Other plants since that time have been built and have been based on the same design.

---

[123] See *Electrical World,* Dec. 31, 1927. Also see "Fleiluft-Wasserdraftanlage," *Eletrotechnik und Meschinenban,* March 26, 1928.

For some time, Joseph had tried to convince the power authority to string long-distance lines in longer spans. When he did the original construction of the Stockbridge lighting system, 100 feet was considered the longest span practical. There was a distance of 250 feet across the Housatonic River in Glendale at the bridge. Joseph had strung a five-sixteenths steel, stranded messenger cable from crossarms. The copper conductor wires were suspended from this by porcelain insulators every ten feet. On one sleety, wintry day, the crossarm at the upper end of the bridge broke. Not only did the #4 insulated copper wires, medium drawn, support themselves, they also supported the five-sixteenths steel messenger cable, the crossarm, and the weight of a half-inch of ice. Joseph pulled the messenger wires off, and the span never gave anymore trouble.

From then on, he advocated and built longer span lines. He also didn't see any reason for insulated wires.[124] If bare wire was good for the higher voltages, it should be okay for lower voltages. Just because a wire was insulated did not mean it was safer to handle. All wires had to be treated as bare conductors. The insulation came off eventually due to ice storms or exposure to temperature changes. When it did, it hung in unsightly strings from the wires. Bare wire collected less sleet, exerted less wind stress on crossarm pins and poles, and lasted just as long if not longer. Finally, and not least, it cost less. Joseph built bare wire lines with spans up to 360 feet. Less fixed charges on the distribution system should have meant lower rates for customers.

In 1930, when he built the Market block, Joseph used a bare wire for the grounded wire on 110/220 volt wiring system. This still was not permitted. Bare wire could only be used for service entrance wires. Joseph could never understand why a wire that was grounded should be insulated. It made a perfect grounding system. It saved copper, the cost of insulation, and space in conduits. At meetings held by Western Massachusetts Electric System in the Pittsfield Electric auditorium, there were discussions of the underwriter's code. Joseph argued for his technique several times, but he didn't receive any results.

Another innovation initiated by Joseph was building distribution lines with bracket construction in place of spread-out crossarms. This method was standard practice in Europe. W. R. Stanley had built the same way in Housatonic

---

[124] Joseph Franz, "Shall We Depart from Present Accustomed Methods of Using Insulated wires on Overhead Lines?" *Parker Public Service Bulletin*, July 1927.

earlier. However, it was not accepted as standard in the United States.[125] The advantages were less cost and less trimming of trees. Plus, it was stronger.

Then there was a constant struggle with the telephone company. They would not go joint[126] in constructing their lines. Their standard spacing of poles was 100 feet with crossarms. Often, they went on the other side of the road to avoid the power company's "outrageous" practice. On a line to Monterey, they set a lower pole halfway between the power line poles because the local selectmen would not allow two separate pole lines. They did change their minds in time, and they now share on long spans with bracket construction.

Always seeking to improve the power plants, Joseph devised a scheme to keep water from freezing at the flashboards[127] in water-powered plants, like the one in Williamsville. He ran a one-half inch pipe about four feet underwater, inserted one-sixteenth inch sprayer nozzles every three feet, and blew air out. A small air pump consuming about six kilowatt hours per day furnished the air. This saved the expense of the men who had to daily cut a channel in the waterway during the freezing weather.

In the stock market crash of 1929, like most Americans, Joseph's career was affected. Joseph had invested in stocks, bonds, and bank saving accounts. He, like all others, lost heavily. It was more than a whole year's income. Luckily, his house and other properties were fully paid off, so the commercial rentals could sustain his expenses.

The C. D. Parker Company fell apart. Mr. Bowen Tufts committed suicide. One trustee after another resigned, and New England Power men took their place. This was completed with ease because the board of trustees was self-perpetuating. In case of a death or resignation, the rest of the board would appoint a member to fill the vacancy because New England Power had a growing majority of trustees who took over. That was a heavy blow to the C. D. Parker interests. The New England Power Company moved into town with Mr. Dustin of Northhampton as general manager. Everyone was under him, and he was under the president, Mr. Bell, in Boston. Obviously, Joseph's time with the

---

[125] It was eventually accepted, and it is used now on long spans for rural lines.
[126] The term used when two or more utilities share the same poles.
[127] Joseph Franz, "Preventing Ice on Flashboards," *Electric World*, December 4, 1926.

New England Power was limited. For one thing, an annual salary of $6,000 during the Great Depression was not for some local yokel. That much money would go to a favorite son. Carl Bidwell was appointed general manager of Southern Berkshire Power and Electric Company.

It was hard for the layman to understand why the big interests wished to make such strenuous efforts to acquire all of these small companies. The Public Service Commission supposedly controlled all utilities. Income was limited. If there were any large profits, the ruling was always a reduction in rates. Nevertheless, Joseph soon learned one could not beat the big corporations. There was always a way to gouge the public.

Granted, things wear out and must be replaced. The transmission line to Lenox had to be rebuilt in 1925. Joseph added a second circuit of three #14 wires to the existing #6 wires. This was completed without a service interruption. The cost was $2,000 per mile. The original cost had been $1,000 per mile. The New England Engineering and Contracting Company rebuilt it again in the 1930s. That time, they used the same number of poles, A-frame type, but they were spaced twice the distance apart at a cost of $17,000 per mile. Engineering alone was $3,000 per mile.

The Public Service Commission had apparently very little control over utility expenses. They once investigated a tower transmission line that was built from Turners Falls to Pittsfield by the Western Massachusetts Electric Company for $27,000 per mile. The Western Massachusetts Electric Company still had its own service company. Even though bids were theoretically let out, their own company always came in low.

When the commission questioned the cost as being too high, the manager simply responded, "That's what it cost. What are you going to do about it?"

That was the end of it. The same sort of thing occurred with the telephone company and Western Massachusetts Electric Company.

Occasionally, someone would call attention to the questionable dealings of the big utilities. Politicians also kept the pot boiling and would point out the discrepancies to their constituents, but the storm always blew over. The people at the top were rarely known. Despite being carefully regulated, the utility industry was obviously fertile ground where big money could still be made.

Joseph's ingenuity finally collided with the power company management over a maintenance job at the Rockdale Plant. Fourteen wooden A-frames supported the steel penstock. The bottom of the vertical support showed signs of decay. The New England engineers were called in to draw up specifications to remedy this defect. Several men came to the plant, took pictures of the supports, drew up plans, and submitted them to the Southern Berkshire Power and Electric Company. Mr. Dustin,[128] Manager Bidwell, and Joseph reviewed the proposal. It called for an expenditure of $6,000.

Not knowing the intricate workings of these high financiers, Joseph spoke, "Hell! I can do this job for $500 or less."

Mr. Duston responded, "If you think you can, then go ahead and do it!"

That is exactly what Joseph did.

He went to the junkyard in Pittsfield, which his friend, Mr. Adelsheim, owned, and bought six-inch L-irons. At the plant, cuts were made partway through the legs. Then two L-irons were strapped on each leg. After the angles were fastened with through bolts, the cut was finished. The piece of wood was removed. A square frame about ten inches high was placed around each leg and filled with concrete. The ends of the legs on the trestle now sat firmly on the ledge. The wood was two inches off the base. The job was done in less than one week. The total cost, including Joseph's time for engineering and supervision, was $450. This must have created quite a commotion in the parent company. The engineering company had already accumulated a couple thousand dollars for their consultant services, and this little squirt came along to upset their beautiful apple cart. There wasn't any need for Joseph's services any longer. He was just a thorn in their side.

By the time the New England Power took over in 1929, Charley Hull had passed out of the picture. After several heart attacks, he had to be confined to bed. One final attack finished him.

As time passed, Joseph continued to be the power engineer, but he was growing more disillusioned. He knew what needed to be done, and he had the expertise to accomplish it. Things were very different now. No project could be undertaken without approval of the company president, the board of directors, and the accounting firm of C. S. Stanwood. Days and weeks could pass before any action was taken. Procrastination was not Joseph's cup of tea. At the company expense, there were many trips to Boston, including golf, wining and dining, and the usual pep talks. Joseph found all of this to be boring.

---

[128] By then, he was a vice president.

In July 1931, Carl Bidwell ordered Joseph to work in Great Barrington. His office would be at the Monument Mill Powerhouse. After twenty-six years of working for this company, this was his reward. Joseph resigned. For the many years of loyal service, he was given a pension of fifty dollars a month, which was later increased to sixty dollars and the benefits of Blue Cross/Blue Shield medical insurance. Family benefits were added after World War II. He was initially told this fee was a retainer for being a consultant, but it was rare when Joseph would be called.

The citizens of Stockbridge once approached him to head a committee to investigate the rates that were charged. Customers on the underground lines paid a higher rate than those served from overhead lines. The company wanted to eliminate all underground service. They hoped pressure from the customers themselves would support overhead lines. On an overhead system, there was much more work for the engineers. Every ice storm brought the lines down. Joseph declined to have any part in this investigation. His sixty dollars per month was far greater than what he might have gained with a lower electric rate. Fortunately, the New Electric Company wasn't a match for the Berkshire residents who refused to have overhead lines in their towns.

Another case was the installation of the tall streetlights in Stockbridge and Lenox. They were placed in the foliage of the majestic shade trees and were of absolutely no value as street lighting. Again, Joseph kept silent. He was glad he had retained the privilege of working on projects of his own.

The price of progress is very high. In retrospect, Joseph would not have changed places with any of the men who now ran the power industry.

Chapter Sixteen

Family and Friends

Returning to Stockbridge in August 1925, Emilia and Joseph had to face the reality of integrating a new person into the Franz family life. When they were married, Natalie had just turned nineteen, and Russell was seventeen. Naturally, both of them initially resented Emilia. Natalie was particularly contemptible. She went for days without talking to Emilia. She just could not see anyone taking her mother's place, particularly because she seemed to think of herself as the present female head of the household. Both siblings had things pretty much their own way for a couple years and thought they couldn't do any wrong. It took a while before peace was restored in the household. Emilia understood their hostility, had a lot of patience, and did not try to provoke confrontations. Natalie graduated from high school. In the fall of 1926, she went to Beaver College in Pennsylvania. Russell was easier to manage alone because he was always eager to please.

Neither Natalie nor Russell had much concern about money. Their father was very strict about requiring an accounting for every penny that was given out. It was a practical way of applying some of the mathematics they were learning in school to daily life. They knew they would have to earn their own way some day, and they knew they should know how to handle their money properly. Joseph was very proud of Natalie when she finished college. She had written for more money to see her through until she got into St. Luke's Hospital nursing school. When Joseph sent her the check, he had once again asked for an accounting for the expenditures.

She wrote back, "I am sending back your check because I believe I am now old enough to be accountable for my own expenses."

That was the last time she ever asked her father for money.

Emilia and Joseph did more surveying jobs together. They finished the map for the golf course and surveyed several private properties, including the Rockwood estate on Ice Glen Road. That was a challenge because part of the property was in the woods on the hill and into the Ice Glen itself. Fortunately, it was past mosquito season, so the woods were delightful.

More news arrived from the family in Europe. Joseph's sister, Barbara (Wetty), and her two daughters made a trip to Vienna in the fall of 1925, shortly after Emilia and Joseph had returned. Their father was failing in health, but he was happy to see everyone. Barbara felt uncomfortable with his second wife and children, even though they tried to make a good impression. Alma was also very ill, so visits were very short. Rosa had just joined the Nazi party, sold her house, and donated the little money she had to the party. In return, she had been given a tiny apartment. She tried getting money from Wetty, the same as she had from Joseph. Wetty did not get to see her sister Marie in Bratislava, but Ludwig and Pepi reported she was doing fairly well. Wetty was saddened to see family members aging poorly, and she was discouraged about the future of Europe. Only Ludwig seemed the same. Her beautiful memories "bei uns in Wien"[129] no longer existed.

Joseph always said that life, in the fullest sense, did not begin for him until he was forty. Until then, he had been focused on having a reputable career and making a good living. The pleasures he had derived were mainly from the joy of life with his family. After forty, he was financially more secure, had built a permanent home, and raised a family. He had weathered a tragic loss and found a new relationship, which promised many rewarding adventures with a new family. He could then indulge in more sports, cultural activities. and community affairs.

When Russell graduated from high school in 1927, he thought he only had to ask for a job to get one. He soon learned that people, who had said they would help him when he was still a child, were not so eager to do so when he did not have any skills or experience. He once went to New York City to see the father of one of his friends.

It surprised Russell when he was told, "Come see me when you finish college."

In the fall, Russell enrolled at Duke University. That only lasted one year because he really did not study. After failing his second semester, he quit. His personality and temperament eventually paid off when he found a good sales job with the Norton Tool and Dye Company in Worcester, Massachusetts.

He was not that far away, so he could join the family in Stockbridge for holidays. He would sometimes even come home for a weekend. On one visit, he showed off his new convertible car with the rumble seat in the rear, in which he

---

[129] "with us in Vienna"

proudly gave rides to his younger half-siblings. He also still loved skiing, resulting in winter reunions on the ski slopes in Western Massachusetts or Vermont.

In 1927, the year Lindbergh flew the *Spirit of St. Louis* nonstop across the Atlantic Ocean, Joseph's family increased with the birth of a daughter, Joanna. Like his father before him, Joseph had happily started a second family. Because Joanna was born in December, Joseph fitted the old sled with a sleigh box so he could pull the little girl on the snow. She loved that small sleigh and the snow. She laughed when a squirrel sat on the edge of her sleigh. A squirrel would sometimes even eat a peanut from her hand. When she started to walk, she would go right to her furry-tailed friends and still offer them peanuts or other food.

Mrs. Sprague Coolidge had just established Music on the Mountain in Pittsfield. Before she was two, Joanna was taken to her first chamber music concert that summer. She did not remember much about the concert, but she never forgot the deconsecrated white church on the hill and sitting very still on the stiff-backed pews while the music was playing.

Natalie became a graduate nurse in 1928. She worked at St. Luke's Hospital, near Colombia University, in New York City. Within a couple years, she married Arthur T. Hewlett, who had been one of her patients. They lived in Hewlett, Long Island, and had one daughter, Anna. Family members would visit them on trips to New York City, and they occasionally visited Stockbridge.

The Christmas of 1929 was the last time Joseph's first family and second family celebrated the holiday all together. By the 1930s, Joseph's older children were adults, away and living independently.

Two months later, in February, while Emilia was in the hospital having their second daughter, Shirley, Joseph took Joanna for a slide down the Laurel Hill path that led into the schoolyard. He sat on the front of the sled to steer it. A large rock was near the foot of the path. Trying to avoid hitting the rock, he stuck out his right foot. The pain was intense as his foot hit the rock. Of course, it was the same foot that had been injured so many times when he was younger. He thought it was broken, but he managed to hobble the short distance home. After applying ice to his ankle to keep the swelling down, he determined it was not broken. It was just a bad sprain. It was traumatic enough that Joanna even remembered the incident.

Shirley was as blonde as Joanna was dark. She was also more adventurous as a small child. Perhaps her parents were less attentive, having to divide their time between two children. Shirley sometimes ventured forth on her own.

Around the age of two, she even went next door to the grocery store. Attracted by the colorful bin of apples, she decided to try one. She had consumed about half of it when her father discovered her.

"Shirley, what are you eating?" he inquired.

The apple swiftly disappeared behind her back with the explanation, "Aw gone!"

Joseph could not help but laugh. He paid for the apple and marched his daughter back home, scolding her the entire way. She finished the apple!

One fall weekend in 1931, Emilia and Joseph went to Connecticut to visit his brother Karl and his family. Karl and his wife, Mary, had purchased a small feed and grain store in Milford. The Franz children were left in the reliable care of Lucy Beacco, who lived with them while Shirley and Joanna were little. Karl's daughter, Teddy, had married Joe Nolan when they were both working in New York City. Joe had been promoted to vice president of the RKO Studio in California. Teddy and her husband had come east for a visit from Los Angeles. It was one of their rare family reunions.

After a nice visit and dinner, they started home. Emilia was driving. It began snowing as they headed north. Going slowly uphill in Goshen, Connecticut, they hit a patch of ice. The car spun completely around and almost tipped over on the right embankment. The passenger side door flew open, throwing Joseph out and Emilia on top of him. Seat belts were unknown then. Thankfully, the door held the old Franklin from toppling all the way over. The frightened couple managed to scramble out of harm's way. They were too shocked to assess their own injuries. A house just across the road had a light on, but, when they approached the door, the light had gone out. They did not want to disturb the occupants. Still in shock, they wandered down the street to the next house, which had a sign on the front lawn that simply read "Guests." They stopped and rang the doorbell. They must have looked fairly bad. Emilia had lost her hat and glasses. Their hair was wet from the snow, and their clothes were wet and askew.

"May we use your telephone?" Emilia asked. "We've just had an accident, and we must find somewhere to spend the night."

The lady who answered the door looked them over and finally said, "Well, you can spend the night here."

She offered them something to eat, but they only wanted something hot. After some tea, they went right to bed. Joseph had hurt his shin, and Emilia had bruised her ribs and elbows. During the night, their bruises began to ache. They had not been thinking very clearly and did not have enough sense to put snow or ice on their injuries before they went to bed. The next day, they managed to get the car towed out of the ditch and made it back home. It took two weeks before their aches and pains subsided. The Franklin fared better. It only had a slightly sprung door.

Later that winter, there was a large snowstorm. Joseph made a snow house for the girls. Inside, it had a seat and a window made of a sheet of ice. The walls were very thick, so he made a slide on one side and laid a stepladder against the other side to get to the top. He sprinkled the whole thing with water that froze. The girls and friends enjoyed it for nearly two weeks.

Figure 29—The snow house, with Joanna on top

Joseph always enjoyed building things, whether it was a large structure or a small one. He built a one-room playhouse in 1931. It was large enough for adults to stand in. Double casement windows were on each side. Inside, there was a loft with a small window near the peak of the roof and a ladder leading up to it. An electric light hung in the center of the room so the children (and sometimes one of their friends) could sleep there in the summer. On the front, a small porch with steps led up to the door that could be locked.

The same year, President Franklin D. Roosevelt took office. Joseph and Emilia's son, Peter, completed the family in October 1933.

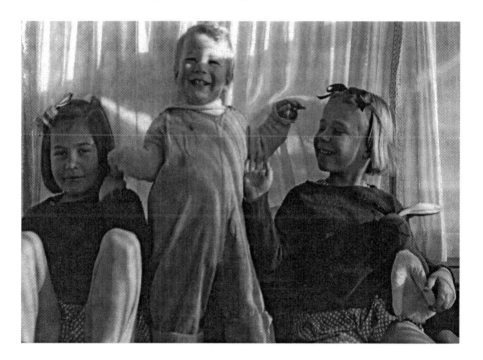

Figure 30—The children of Joseph and Emilia, Joanna, Peter, and Shirley

Roosevelt's New Deal policy promised the return of financial stability, but that would not happen overnight. Meanwhile, the Franz family practiced austerity and frugality for the next several years.

So many others were worse off than the Franz family. Desperate men would come to their house asking for any kind of work, just so they could have a meal. There was a lawn to be mowed, snow to be shoveled, or windows to be washed. There would always be a hot meal as well as some small compensation for their efforts. No one was ever turned away, even if no work was necessary. They always packed fruit and sandwiches when they sent the men on their way.

One fun project was building the dollhouse that Emilia decorated for a Christmas present for the children in 1937. It was a challenge because they had to work at night after the children were asleep. The dollhouse was two-sided. Both front and back walls were removable. A staircase complete with spindles and a curved banister went between floors in the central hall. The living room had a fireplace with andirons and a complete set of metal fire tools that Joseph had handmade. On the first floor were the living room, dining room, kitchen, and hall. The second floor had four bedrooms, a bath, and hall. It was furnished in typical 1930s style. The windows opened like the double-hung sash type, and curtains hung on every one. Each room had its own electric light illumination. Additionally, Joseph made the front door with its own lock and key. The back door had a hasp lock on the inside.

Creating useful objects was another hobby. When he built the house, Joseph designed and constructed the outdoor lantern that hung on the outside corner of the office wall. He made his own tools for specific jobs when standard tools did not work. All of these things were examples of his artistic ability.

Emilia and Joseph were both creative people, so they supported each other in their particular interests and shared family responsibilities. Emilia often did the lettering for Joseph's blueprints and checked the spelling on letters and articles. Both were rugged individualists, believing in self-sufficiency and the right to develop to one's full potential. They understood that, the more interests and skills one had, the more equipped one would be to accept new opportunities.

During Joseph's last years with the power company and after he retired in 1934, he had more time to be with his children. Though relatively strict disciplinarians, as parents, they taught their children to view the world with open minds and open hearts. Of prime importance were the responsibilities of belonging to a family and community. Yet, they encouraged limitless goals within a structured environment. The family participated in many physical and cultural activities. Joseph took over the children's sports education while Emilia tended to the academic and cultural side.

Peter was a handful while growing up. He was very active and curious. He loved taking things apart, even though he rarely could put them back together again. His curiosity cost him the loss of part of his right forefinger when he stuck his finger into a cogged gear of a nonworking elevator in a friend's attic. Cuts and scrapes were the badge of his reckless enthusiasm and the source of much of his mother's anxiety.

Sex education began at home in a very natural way. Emilia wanted her children to be better informed than she had been. Her mother had only told her that babies came from cabbages. Procreation was a part of life, as was birth and death. If there were questions about their bodies or where babies came from, they were answered candidly and correctly. Based on the child's age, as much information that seemed warranted was provided.

In the hot weather, the children shed their clothes in the backyard. They swam naked in a gentle rainstorm or under the outdoor shower that Joseph had installed near the back of the house. Inside, there was always an open-door policy for the bathroom. The youngsters were allowed to use the toilet, even if their parents were bathing or shaving. Joseph's straight razor fascinated the youngsters. Nursing the baby was not hidden from the older two. Conversations of any nature never stopped just because a child was in the room.

As the children grew older, they were instructed in a number of practical tasks. They were given regular lessons in the basement where Joseph taught the proper use and maintenance of all kinds of tools. Specifically, they were taught how to sharpen a chisel or draw knife or plane, the differences between types of screwdrivers, the differences between a crosscut, rip, and hacksaw, and the differences between a pipe wrench and a monkey wrench. The children learned how to wire a lamp; turn off the electricity, gas, and water; change a washer; or fix a broken lock. Upstairs, Emilia trained them in basic housekeeping skills, including cleaning, making beds, doing laundry, sewing, washing dishes, setting the table, simple cooking, and baking. Each child had his or her specialty in baking. Joanna was pies. Shirley was cakes. Peter was gingerbread.

"You never know where life will lead. You may be fortunate to either make enough money to afford servants or marry a rich person. If you don't know how to do a job yourself, you won't be able to tell anyone else how to do it, and you'll never know if the job is done properly."

This was the explanation for all of these activities.

Winters were filled with after-school skiing and skating. Sometime in the 1930s, Joseph had switched from his racing skates to figure skates. He had never been competitive in any sport, but he liked to do well at whatever he tried. He learned some of the primary figures and a few simple dance steps. The children also learned to figure skate. It wasn't the fancy stuff one sees today, but it was the school figures and some ice dancing. Emilia still skied and skated a little, but keeping up her own social schedule and running the house didn't give her much time for outdoor activities.

Horseback riding and swimming were the sports of summer. Joseph would rent a horse from Ostrander's Stable on Goodrich Street. He would sit a child, barely able to walk, in front of him on the horse and walk around their yard behind the Block. When the children were old enough to ride by themselves, ponies or horses were hired for each one. First, they rode in the paddock until they learned to handle the horse and ride in proper English style.[130] When they gained enough experience, Joseph and his children took trail rides on the roads in the area that were still unpaved. Several people in and around Stockbridge owned horses. At times, they rode with a group of three or four. Joseph's children were often embarrassed when he started singing, "Home on the Range," but no one else seemed to mind.

One of the riding compagnions was Engel Krichels, head of the Hurlbut Paper Company in South Lee. Joseph met him when he had done some work for the mill. He came from Frankfurt am Main in Germany, so they could share their common heritage. Engel had ridden in European competitions as a young man and still loved the sport. He also was an excellent skier. His sister, Tückie, though not interested in sports, became a good friend of the family as well. She had a great sense of humor and an infectious laugh.

Count Malevsky was from Russia and lived in West Stockbridge. He gave everyone pointers to improve their riding skills. He explained, in the Russian cavalry, one had to put a ruble of your own pay between each knee and the saddle. If you dropped the rubles, you did not get them back. You learned to stick to the horse very fast.

The Berle family also rode and kept horses. Joseph first met Dr. Beatrice Berle when both of them had served on the Monument Mountain Reservation Committee in the 1920s. She and her husband, Dr. Adolf A. Berle[131] had a house and farm on Muddy Brook Road in Great Barrington, where they lived

---

[130] Everyone rode English style in the East.

[131] Dr. Adolf A. Berle was ambassador to Brazil and later assistant secretary of state under Franklin D. Roosevelt.

with their three children. When the Berle family were away in the fall and winter, Joseph and the children exercised their horses and pony. They had several fields and dirt roads on which to ride. They also learned to saddle a horse and to curry comb and brush the animals after their rides. It was like having their own stable.

The Berle family had once driven a horse-drawn carriage, a Brougham. It was an elegant vehicle with all-leather interior, beveled glass windows that could be opened in three different positions, lovely lanterns, a leather-covered coachman's seat, ivory knobs on the door handles, and other attachments. It was abandoned in one of their fields.

Joseph asked if they were using it for something special. Mrs. Berle laughed and said, "Heavens, no! If you want it, you are welcome to haul it away."

They were glad to get rid of it. Joseph hitched it to his car and brought it home. For many years, it stood in the back lot, near the playhouse, where the children and their friends had many happy adventures. It was a sad day when the wood rotted away and it had to be dismantled.

Charley Ackley and his family still lived in Interlaken, across from the church on the hill. They also maintained their small lot on the Stockbridge Bowl,[132] right next to the Stockbridge town beach. Their son was grown and away in the diplomatic service, so Charley and his wife only used it occasionally. The Franz family continued to have permission to use the place as much as they wanted. Joseph had made several improvements to the place, even though the toilet facilities were in the same primitive outhouse in the woods. The main cabin opened up on the front, which faced the lake. The large barn doors swung out, and an outside deck extended the interior another few feet. An old, woodburning cookstove supplied heat on chilly or rainy days. Mainly, the cabin was for storing the picnic table, a rowboat Joseph had purchased, and a sailing canoe. The Franz family often had picnics there. Sometimes, they were alone. Sometimes, they were with the Ackleys and other mutual friends.

Joanna took to the water like a fish. She swam when she was three. Shirley and Peter took a little longer. When they swam well enough, Joseph and his children would swim to the second raft that the town maintained for diving. Joseph did not dive because he feared throwing out his right shoulder again, but the

---

[132] This lake has two names. At the south end, it is called Stockbridge Bowl. At the north end, it is known as Lake Makeenac, the original Indian name.

children learned to dive from the board and, later, the tower. The first raft, which was close to shore, had a slide for those who were not strong swimmers.

Joseph would occasionally take out the rowboat so the girls could learn to use the oars and steer it. Later, all of the children learned to paddle the canoe. Just in case, they even learned to tip the canoe over and right it again. The canoe was lighter and easier to handle, so that was their choice most of the time. On the western side of the island, there was a shallow cove where water lilies grew. When the flowers were blossoming, they would paddle there and pick a bouquet of white and yellow flowers for their mother. When they could not go to the lake, the children enjoyed playing in the backyard under the outdoor shower.

When the girls became teenagers, they were given tennis lessons,[133] even though their parents did not play the game. Peter learned in high school at Deerfield. Joanna and Peter added this sport to their list of lifetime activities.

Emilia was adventurous and loved to travel. In their early years together, they spent at least one or two weeks each summer at the seashore or a lake. They would rent a house. Sometimes, they were with another couple. Sometimes, they went alone. In 1929, they hired Miss Killburn, a registered nurse, to look after Joanna so the couple could go to Atlantic City for a week. In the fall of 1931, they visited Lake George in New York State. In September 1932, when the girls were old enough, they shared a cottage with the Ackley family in Guilford Point, Connecticut. In July 1933, they were in Saybrook, Connecticut. The summer of 1934, Emilia, the three children, and a pet robin Emilia had nursed back to health spent three weeks in South Hero on Lake Champlain. Peter was not yet one. Joseph joined them for the final week. Sadly, the robin died. The children held an elaborate funeral service on the beach. The family procession led to the robin's final resting place, where Joseph had dug the grave.

Train travel continued to be the connecting link between both Joseph's and Emilia's extended families. The New York, New Haven, and Hartford Railroad ran from Pittsfield to New York City, stopping in many of the small towns along the Housatonic River along the way. The family took the four-hour trip once or twice a year together. They stayed at the Commodore Hotel,[134] centrally located next to Grand Central Station on 42nd Street.

---

[133] Peter learned how to play in prep school.

[134] It is now the Grand Hyatt.

Both parents had relatives who were still in the city. Luise and her sister, Anna, lived together in the Bronx after their husbands died. Wetty and her oldest daughter also lived in the Bronx. Emilia had an aunt and two uncles in Brooklyn. Aside from visiting relatives, the family enjoyed seeing a show on Broadway and trying new foods in a nice restaurant. Joseph and the children also enjoyed ice-skating in Rockefeller Plaza and at the Gay Blades rink on top of Madison Square Garden, which was then located on 49th Street. They bought new ice skates or new boots when they outgrew the old ones. They also attended the ice shows at the Center Theatre with Sonja Heinie and comedians, Frick and Frack, with their forced edge, spread-eagle backward. Peter learned how to do this maneuver when he was in college. Joseph would sometimes go alone or with Charley Ackley. Charley and Joseph also stayed at the Commodore. They usually managed to catch a show at Minsky's Burlesque on 14th Street after a dinner at Luchow's.[135]

In 1936, LaGuardia Airport opened, starting cross-country passenger flights. Shortly after it opened, the Franz family visited the new airport. It was exciting to stand on the balcony overlooking the water and watch the aquaplanes land in the East River. The children knew their Cousin Teddy and her husband had actually flown from California to New York City on one of them.

The summer of 1937, Emilia took the kids to Los Angeles by train. Before continuing, they spent one night in Denver to see Emilia's relatives. In Hollywood, the family stayed with Joseph's niece, Teddy. Because her husband, Joseph Nolan, worked at the RKO Studio, they lived in a lovely apartment near the studio on Rossmore Street. On the way home, Emilia and the children took the train up the West Coast and across Canada to Lake Louise and Banff. Joseph was unable to go with the family that time, but he was happy they had such a pleasant, educational experience.

Of course, as with any family, they sometimes had problems, including various accidents and illnesses. Joanna was once going to a classmate's home after school. While waiting for the school bus for the run to Cherry Hill Road, the children were playing on the swings. Joanna was pushing her friend and looked away just when the swing came back and hit her head. It knocked her out, and the school nurse was summoned to the scene. She brought Joanna to her office and called home for someone to come get her. Joseph rushed to the school. Joanna was bleeding rather badly from a gash near her eye. He carried

---

[135] It was a famous German restaurant that stood on 14th Street, near 4th Avenue South.

her home, cleaned her up, taped the gash together tightly, and gave her an ice pack to bring down the swelling. The next day, she went to school with an obvious black eye. In due time, the cut healed, leaving only a small scar.

With all of the children in school, they were exposed to the usual childhood diseases. First, measles and German measles made the rounds. In the fall of 1937, the kids had just gone through whooping cough when chicken pox started. Emilia was tired of playing nurse and decided the family would avoid the disease this time. In January 1938, on the advice of Dr. Hunt, the family pediatrician, she took the children to Florida for two months. As it turned out, the children had already been exposed, so the chicken pox followed them south. The following summer, when Peter contracted scarlet fever, the whole family was quarantined for six weeks. They were not able to go anywhere. The vaccines for all of these diseases were not developed until after World War II.

Joseph and Emilia enjoyed a full social life in Stockbridge. Emilia was a founding member of the ladies' Tuesday Club. Both she and Joseph were active with the Berkshire Garden Center[136] and the Laurel Hill Association. Their closest friends were probably Hap and Roxie Davenport as well as Flo and Karl Etting. Hap was a school principal. First, he worked in Stockbridge. Later, he worked in West Stockbridge. Emilia and Roxie had been good friends before she was married. Joseph had known Karl Etting, an architect and immigrant from Denmark. They shared similar experiences in adjusting to their adopted country. Sadly, Flo and Karl split up and moved to New York City. Flo and her daughter, Flolydia, ended up in Chicago. The family saw them a few times in the New York City and, later, en route to California or on the way back.

Because Emilia was born and raised in Stockbridge, she knew many of the inhabitants very well. Most of their friends were other business people in town. There was Ethel and John Palmer, a banker. Edith and Henry Derrick ran a soda fountain and gift shop in the Guerrieri block. They also knew Charlotte and Ned Wilcox. Ned worked for Charley Hull. This group, Jane and Charley Ackley, and the Davenport family comprised their bridge and golf partners. Emilia was a good hostess and liked entertaining. Many weekends, from the fall through the spring, there were bridge parties at one house or another. Desert and coffee were served. Sometimes, dinner was served. Through their children's activities, they met some of the younger Riggs doctors and their fami-

---

[136] It is now the Berkshire Botanical Gardens.

lies. Dr. Hiden and his wife, Tilly, shared many similar interests. Their son Bobby and Peter remained lifelong friends.

Joseph maintained a longtime friendship with Sol Adelheim, the junk dealer, from Pittsfield. Over the years, Joseph had purchased many parts from him. Being Jewish, Sol had a hard time with his Yankee neighbors and customers. Everyone tried to get the lowest possible price for goods they bought and then would laugh at his attempts to please them. Joseph could sympathize with his frustrations. Sol was well-educated and could discuss any subject with knowledge and understanding. Joseph did not see him often, but Sol would always make a point of bringing the family a fresh baked challa[137] for the holidays, both Jewish and Christian.

In 1939, the first World's Fair opened in Flushing Meadow, Queens, New York City. That summer, the family went to the city to see it. There were too many pavilions to see in one visit, so, the following year, they went a second time during the children's spring vacation. New York City had built a special spur on the elevated Flushing subway line to get from 42nd Street to the fair. One could get there in no time at all. The Trylon and Perisphere, symbols of the fair, welcomed people from all over the country and overseas. The engineering of both amazed Joseph. To Peter, they were just a big ball and a pointy stick. Joseph particularly enjoyed the Westinghouse exhibit with its huge automated figure of "Ready Killowatt" and the General Motors' "World of Tomorrow."

It was good to see that people still had the imagination and ability to venture into the future with positive new ideas.

---

[137] a sweet bread

Chapter Seventeen

Community Commitment

When Joseph paid off his mortgages in 1924, he began investing some money in the stock market. He was doing fairly well until the crash in 1929. Real estate seemed a more secure investment. The Block was bringing in a fair amount of rent from the telephone company upstairs and the post office downstairs. Mr. Monroe, the original tenant on the first floor, had moved his bicycle shop to the small rear office. Joseph thought people in town would always need to buy food. In 1930, he designed and built the market building between the Block and the house. It was constructed of hollow tile blocks and covered with stucco. It had a flat roof made of tar and gravel. The house was painted a pale yellow while both business buildings were left the natural gray of the cement stucco. All buildings were originally heated by a boiler in the basement of the Block. A driveway between the two buildings gave access to the parking lot and Joseph's garage.[138]

The A & P grocery store was the first tenant, the second chain grocery store in town. The other was the First National Store. Ironically, neither of these stores survived the Great Depression, but smaller, locally owned groceries took their places. Joseph's building has always housed a grocery store of some kind.[139]

In addition to building, Joseph loved the freedom of being out in nature since childhood. Among the trees, lakes, and rivers or high on a mountain or hill, he could air his thoughts and dreams without fear of ridicule or scorn that he had often faced in his work or trying to establish a place for himself and family in the community. The trees and rocks did not care if he was foreign-born. Numerous state parks and reservations in the Berkshires afforded him many hours of quiet contemplation.

---

[138] The garage was torn down in 1996 when the block was sold.

[139] It is currently known as the Elm Street Market.

One of these is the Monument Mountain State Reservation between Stockbridge and Great Barrington. On the south side, it has granite outcropping, which people thought resembled the head of an Indian chief, hence the name. It also has its own local Native American fable.

"Long ago, when the Mohican tribe still hunted in these hills, the chief's daughter fell in love with a brave, whom her father did not like. He forbade his daughter to marry the man. Their love could not be quelled, so the desperate pair threw themselves off the mountain's south cliff. Instead of living apart they died together."

Over the years, trees have grown too tall to see the granite outcropping or even the cliff, but the spirits of the dead lovers perhaps still linger.

Hiking the trails on Monument Mountain had been one of Joseph's first pleasures in the Berkshires. When he finally moved to Stockbridge, he became a member of the local committee that oversaw the management of this property. In the early 1930s, when Chairman Bernard Hoffman resigned, Joseph was elected chairman. The committee included several Berkshire residents. The Stockbridge residents, in addition to Joseph, included Alice Riggs, Rodney Proctor, John Butler Swann, and Edward F. Belches. The Great Barrington residents included Frank Pope and Dr. Beatrice Berle.[140] The Monterey resident was Earl Stafford. The Sheffield resident was Walter Prichard Eaton. Arthur C. Monroe, from Stockbridge, was the reservation warden. Some of the committee's accomplishments included building and placing new signs when and where needed, maintaining the picnic area near Route 7, eradicating poison ivy from the trails, and mowing the grass in the picnic area and trailheads. Joseph served on the committee more than thirty years, finally leaving in 1948.

Another group Joseph found stimulating was the Laurel Hill Association, the oldest village improvement organization in America. It was founded in 1853 by Mary G. Hopkins to "keep the town neat" and plant and protect the trees in order to preserve the natural beauty of the town of Stockbridge. It also maintained the Indian burial ground, Laurel Hill Park, Ice Glen, and other designated green spaces in town. They sponsored Laurel Hill Day each year to promote increased awareness of nature conservancy and encourage more people to become members. Joseph joined the organization in the early 1920s.

---

[140] Joseph first met the Berle family when he served with Beatrice on this committee.

Regular meetings were held in various locations, including private homes. Several meetings were held in the Franz home over the years. In 1928, Joseph was elected treasurer, an office he held for thirteen years. He and his second wife enjoyed working with many different people in the organization. Some were the wealthy cottage people. Others were townspeople, merchants, craftsmen, teachers, farmers, clergy, and service providers. It was a genuine town-oriented organization that was open to all.

The annual meetings were always held in August in the clearing on Laurel Hill, where there was a small platform with a stone bench and podium. The 1939 meeting was a bicentennial celebration of the incorporation of the town of Stockbridge. It was a pageant of the history of the town, from prehistoric time to the present. It was quite impressive. Emilia and the girls were dressed as Indians. Peter was a giant Japanese beetle. Emilia made the girls' and her costumes. Joseph made the papier-mâché beetle costume for Peter.[141]

The Laurel Hill Association built the original Laura's Tower, a wooden observation tower sitting on the crest of the southern arm of Beartown Mountain above Ice Glen. When it deteriorated, it had to be condemned. Joseph volunteered to design a new steel tower to replace it. It was approved. Money was raised to build a new tower. Joseph ordered the materials, hired the crew, and directed construction. It was completed in the summer of 1934.

Another job Joseph contributed to the Laurel Hill Association was the Ice Glen Bridge. The first concrete bridge across the Housatonic River, built in 1895, Mary Hopkins Goodrich donated. When it was found to be unsafe, it was destroyed with. dynamite. After the Berkshire trolley stopped running, the steel trolley trestle replaced it as a temporary measure. In the 1936 hurricane, the river flooded and washed away the steel bridge. Joseph designed a new steel suspension bridge with stone arches for anchors and supervised its construction.

---

[141] See Walter Prichard Eaton, *Laurel Hill Pageant*, printed for the Laurel Hill Association, August 1941. Also see *Times-Union*, Albany, New York, September 1, 1939. *Berkshire Evening Eagle*, Pittsfield, Massachusetts, September 2 and September 4, 1939

Figure 31—Ice Glen Memorial Bridge

Approximately midway through the work, there was a terrible accident when the workmen were fastening the bridge hangers to the cable.[142] The framework was not yet supported. Vibrations of the men working and the weight of the steel were just too much stress. The entire thing collapsed, throwing four men off. Attillio Gregory was killed instantly when he landed headfirst on a rock. The other three survived with minor injuries. Joseph was devastated. Nothing is more horrible than witnessing an accident and be totally helpless to prevent it. Despite this tragedy, the bridge was finished in the allotted time.[143]

---

[142] "Bridge Buckles Killing Workman in Stockbridge," *Berkshire Evening Eagle*, August 31, 1936.
[143] It still spans the Housatonic River today.

In 1931, Augustus Lukeman, a well-known sculptor who made his home in Stockbridge, was commissioned to create a war memorial in Pittsfield. Joseph had known Mr. Lukeman's work from his previous work in 1905 when Lukeman and his fellow sculptor, Daniel Chester French, created the stone rostrum[144], bench, and desk for the Laurel Hill Society. Mr. Lukeman asked Joseph to survey the triangle of land on which the memorial was to sit. He needed exact measurements to judge the size of marble that would be appropriate for the space.[145]

Joseph believed America had given him opportunities he could never have achieved in his native land. As a retiree of the New England Electric System, he felt he should do something for the advancement of the arts and sciences in return for what had been given to him. His later years were primarily devoted to philanthropic enterprises.

Dr. Henry Hadley was a famous conductor and composer with the New York Philharmonic Orchestra. In the summer of 1934, on one of his many visits to the Berkshires, he conceived the idea of a Berkshire Music Festival. Perhaps the outstanding cultural development of the Western Massachusetts County inspired him. Hardly a town could be mentioned from Mount Greylock in the north to Mount Everett in the south without conjuring up memories of the famous artists, writers, musicians, actors, historians, scientists, politicians, and educators who lived there or visited the Berkshire Hills. Dr. Hadley advanced the idea to Miss Gertrude Robinson Smith, a lady with a dynamic personality and many influential friends. Unlike many of the summer homeowners, she actually did much of the construction work on her Stockbridge summer home herself. First, she consulted Owen Johnson and several others in the area about a music festival. After they responded positively to the idea, a larger group, which included Joseph, was then invited to join when Dr. Hadley publicly presented the plan for a music festival.

The founding officers were Mrs. Elizabeth Sprague Coolidge[146] as honorary president. Other officers included Miss Gertrude Robinson Smith as president. Mrs. Carlos V. de Heredia, Norval H. Busey Jr., Owen Johnson, and Miss Mabel Choate were vice presidents. George Edman was the clerk, and John C. Lynch was the treasurer. Joseph became a member of the board of trustees.

---

[144] platform

[145] The statue can best be seen when leaving the city going south on Route 7, thanks to the suggestion of Emilia, who was also a friend of the artist.

[146] Founder of the South Mountain Chamber Music

Committees were formed in various towns. Meetings were held, and the idea of "Music under the Moon"[147] was successfully promoted to the public as a summer music festival.

Dr. Hadley assembled an orchestra of sixty-five players from the New York Philharmonic Symphony. Joseph designed and supervised construction of a plywood shell that was built in a horse show ring at the Hanna Farm[148] in Interlaken. The farm sloped up from the western shore of Lake Makeenac. The horse ring stood in a flat field on the pinnacle between the house and the barns. To the west, there is a mountain with thick woodlands bordering the property.

Joseph additionally designed the benches that were set in front of the shell. Seating was available for about 2,000 patrons. The benches arrived a day before the first concert, but they were not painted. That was completed in the afternoon and the next morning, but, by noon, the paint was still damp on some. The solution was to cover all of the benches in brown wrapping paper. That solved the wet paint problem, and the paper was removed by the second concert.

The entire community was imbued with the spirit of the festival. Many offered their services for parking, as ushers and ticket takers.[149]

The first series of three concerts were played under a bright, starry, moonlit sky on August 23–25 in 1934. During the first concert, Joseph heard a distinct echo resounding from the nearby mountain. The next day, there was a short rainstorm, typical of the summer mountain squalls. That evening, there wasn't any echo. Joseph realized the wet shell had overcome the echo. From then on, he saw to it that the shell was hosed down before every concert.

Those who listened to the music, superlatively played in such verdant surroundings, must surely have been as emotionally impressed as Joseph was.

The *Berkshire Evening Eagle* reported glowingly, "The acoustic properties of the shell were excellent, all things considered. The music carried to the last bench, which seemed to be 300 feet away, clearly and easily."

---

[147] A booklet with this title, by John G. W. Mahanna, was subsequently published in 1955. It told the history of the Berkshire Symphonic Festival and included appropriate photos.

[148] The Hanna Farm was formerly the home of Dan R. Hanna, son of United States Senator Mark Hanna. It is now the Stockbridge School.

[149] This same spirit has continued over the years and adds to the unique ambiance at the concerts.

Eighteen hundred to two thousand patrons attended each performance of the first festival. It was a financial success. The eight guarantors, who voluntarily backed the event, had to make up only $51.67 apiece. The stage was left in place, but the shell and benches were stored for use the following summer. In the fall of the same year, the Berkshire Symphonic Festival was incorporated under the laws of Massachusetts as a not-for-profit organization. There were uncommonly broad provisions to encourage education in the arts; organize, conduct, and support an annual music festival; provide opportunities to develop musical and artistic talents; and engage in any other activity that would ultimately aid and foster the arts. There was no capital stock, and the corporation was not engaged in the purpose of profit. It was solely to contribute to the development of art and music.

Because the response to the first concerts had been so encouraging, plans were made for a second season at the same location. Dr. Hadley again conducted, and the orchestra was enlarged. The Berkshire Music Association, a chorus of 200 voices, was engaged to augment the instrumental performance for one concert. Rudolph Gans, a pianist, was engaged to perform as a soloist at another concert.

During the first concert of the second year, the weather was threatening. As a precaution, a large circus tent was erected, along with the open-air arrangement of the previous season. This proved a wise expedient when a rainstorm coincided with one concert.

The attendance had now swelled to about 3,000 for each concert, but the festival still faced financial problems. These were met with initiative and imagination. Dr. Hadley resigned because of a prior commitment. Instead of attempting to retrench and give the public less for its money, the management approached the Boston Symphony Orchestra with an inquiry as to whether it was available for the festival's future seasons.

This invitation had an unforeseen interest for its conductor, Dr. Koussevitzky. He had never been pleased with socially stratified audiences, like those in Boston. The plan for the Berkshire Symphonic Festival was similar in certain respects to one he had envisioned in Russia just before the outbreak of the revolution. He also hadn't forgotten the tours with his orchestra that were taken by boat along the Volga River, giving performances at nominal prices for the population. Thus, the famous conductor agreed to bring the Boston Symphony Orchestra to Stockbridge the following year.

Joseph's community activities were influenced by national politics, specifically Franklin Roosevelt's election in 1932. That forever changed the way things were done in America. Joseph was initially skeptical about Roosevelt's prospects as a leader. He soon learned that "business as usual" no longer existed.

For a few years, Joseph entered the political arena himself. In 1935, there was a special election for the Stockbridge Board of Selectmen. Walter Paterson resigned from the board shortly after being elected. His seat had to be filled. For years, Joseph had complained about the ultraconservative tactics used to run the town. He felt they were not in the best interests of the people. Moreover, everything cost too much. It was time for him to do something about the situation, if he could.

In Massachusetts, a group of people known as selectmen govern most of the 281 towns. The number of selectmen is between three and seven members, depending on the size of the town. The citizens of the town elect them. There is always an odd number to prevent deadlock decisions. Stockbridge has always had three. Of those three, one is chosen as the chairman. That person represents the town for any official functions, similar to the duties of a mayor. Generally, Western Massachusetts votes Republican. Therefore, to be elected, there was a better chance if that party endorsed one. However, by often speaking his mind at town meetings, Joseph had shown there were alternative methods of doing things and doing them cheaper. He did not get the Republican endorsement, but he did get the Democratic approval. He ran against Henry V. Rathbun, Republican, and Clark Knight, Workingmen's Party. Dubbed as the "battle of the century,"[150] the campaign was certainly heated. The main issues were the road situation and the police.

Joseph's election was an apparent revolution for Stockbridge as the *Berkshire Evening Eagle* reported that "Joseph Franz bowls over two competitors."[151] In another issue, the headline was TREADWAY'S TOWN GOES DEMOCRATIC.

The board of selectmen consisted of Henry B. Parsons, Nathan M. Shaw Jr., and Joseph.[152] Mr. Parsons was the chairman. In 1935, $10,000 was removed from the town's expenses, which was carried into the next year as surplus. That meant a reduction in taxes. Walter M. Stannard was not rehired as chief of police. Instead, the chairman of selectmen would serve as police chief. At the

---

[150] "Battle of Century Looms as Voters Ponder Choice for Board of Selectmen," *Berkshire Evening Eagle*, March 27, 1935.

[151] "Joseph Franz, Democrat Elected in Stockbridge," *Berkshire Evening Eagle*, April 2, 1935.

[152] Other than Joseph, the other two selectmen were Republicans.

town meeting, a resolution was also passed to appoint one person as road superintendent instead of having three foremen. Thus, the spoils system in the highway department was eliminated. Then the town elected to buy a piece of road-building machinery, a road scraper.

When the salesman came to Joseph's office, he proffered, "The machine costs $800, but there's $100 in that for you."

Joseph countered, "I don't want the $100. Just sell the scraper to us for $700."

Shocked, the man announced, "We can't do that! It would spoil it for everyone else in all the other towns."

Therefore, the town paid the price. Joseph put the $100 into the town treasury. He was not going accept any kind of a bribe.

In 1936, the entire board was up for election. Parsons was the choice of the Republicans. Both Junior Shaw, as he was known, and Joseph ran on the Democratic ticket. They had stepped on too many toes for the Republicans. All three were reelected. Joseph was chosen to be chairman and thereby became the chief of police as well.

Figure 32—As chairman of the board of selectman, Joseph threw out the first ball for the Stockbridge baseball team's home game.

For the thirty-fifth year, Allen T. Treadway was moderator for the annual town meeting. A vote of appreciation for his long service was entered into the minutes. The congressman urged the town to do something about collecting the $60,000 in delinquent taxes to help pay the town's bonded debt of $30,000. A vote passed, requiring the tax collector to keep definite office hours after the tax bills were issued each year. Previously, no hours had ever been designated.

The next year, the selectmen made several significant changes in the town affairs. They cut their own salaries by more than fifty percent because their duties were decreased by appointing Mr. William Healy as road supervisor. Requisitions for all purchases made in the town's name had to be secured before buying anything. Deeds had to be recorded as soon as executed.

As instructed by a letter from the Department of Public Safety, a law was enacted to require a photograph and fingerprints from all applicants for gun licenses. That included Joseph. For his duties as chief of police, he was issued a pistol, even though he rarely carried it due to his abhorrence to any sort of violence.

Stockbridge joined the Housatonic River Association to help clean the river. It was one of eight towns cooperating on this project. The others were Lee, Dalton, Hinsdale, Pittsfield, Lenox, Great Barrington, and Sheffield. Joseph was one of the committee that represented Stockbridge.[153]

During his tenure as a selectman, Joseph made a study of school affairs and found, compared to the average of other towns, that Stockbridge was spending too much per pupil. The Massachusetts State Department of Education in Boston determined the school accounting system. Separate appropriations cover all costs, including the salaries of the superintendent and his staff, building costs, land acquirement, bond payments, interest on bonds, insurance of buildings and equipment, school lunches, and extra expenses like major repairs. Joseph pointed all of this out at a PTA meeting.

Mrs. Edman Wilcox, a former schoolteacher and member of the school committee, complained, "You cannot charge these expenses to the school."

Joseph questioned, "Who pays for it?"

She answered, "The town pays for it."

---

[153] "Housatonic River and Its Future Discussed by Stockbridge Group," *Berkshire Evening Eagle*, October 4, 1936.

At the time the study was made, three pupils in were Latin #2. None in the Agricultural Vocation School were from Stockbridge. The agricultural school had approximately fifteen students from West Stockbridge who paid tuition. Some support came from the state funds. However, it still cost Stockbridge about $3,000, plus the use of rooms for those classes and subsidized lunches for those students.

Many years later, Joseph served on the school board to fill the term of Percy Musgrave,[154] who had resigned. He got no more support then in trying to bring about any improvements. Matters worsened as more valuable property went off the tax list. Anson Phelps Stokes estate went to the Jesuits, a nonprofit religious order. Karl de Gersdorf's mansion burned to the ground. Eden Hill was sold to the Marion Fathers, also a nonprofit religious order. With fewer taxes, there was less money for the schools.[155]

Like many other small communities throughout the United States, Stockbridge has a volunteer Fire Department. The firehouse was on Elm Street, just a few doors from the Franz house. If there was a fire, the men came from all over town. Those who lived the closest to the firehouse got the trucks out. The rest joined them at the site of the fire. Ed Pilling was the chief when Joseph was a selectman. Joseph monitored the response time it took to get the trucks out after the siren sounded. Telephone operators still called the chief and activated the siren. The chief then called the men closest to the station and told them the location of the fire. Everyone else called the operator to also find this information. It sometimes took five minutes to get the trucks out of the station. Joseph began pestering Ed Pilling to see if it could not be done faster.

They got it down to two minutes or less, but, even in the middle of the night, Ed always called Joseph to ask, "Was that quick enough?"

Joseph faced a number of crises as chief of police. He was once on the way to the drugstore when Leslie Stannard came running out of the store with a bottle of gin in his hand. Charley Ackley was right behind him, yelling, "Thief! Stop him!"

Leslie headed down Main Street as Joseph chased after him. He finally caught the man by the shoulder in front of the Martin Inn.[156]

---

[154] His son, Story, became an astronaut.

[155] The towns of Stockbridge, West Stockbridge, Housatonic, and Great Barrington eventually built a regional high school.

[156] The Martin Inn was formerly the Hull's house.

Joseph said, "You're under arrest."

Leslie swung around, punched Joseph on the jaw, and ran down the street. Joseph was too stunned and bleeding to give chase again. The police officers finally picked up the culprit at his home.

At another time, a cow got loose from Robert G. Stewart's Eden Hill Estate. She galloped down the hill onto Main Street, leaving a trail of cars with scratches, dents, and broken windshields. Someone managed to get a rope around her neck, but she broke loose again. A crowd gathered to watch the fun. As Joseph raced up Elm Street, brandishing his gun, he heard people yelling "Mad cow! Mad cow!"

Just as Joseph reached the scene, the wrecker truck from Paul Ruesch's garage arrived. The cow was lassoed again. This time, they got chains from the truck around her. She was finally led back to her own stall, still kicking and mooing behind the truck.

Stockbridge doesn't have any stoplights in town. There is a steep hill where Route 7[157] comes into town. At the intersection of Route 7 and Main Street, there is a stop sign. It is a very dangerous corner as traffic also enters town from East Main Street. Many cars came down the hill too fast and failed to stop at the sign. Joseph, wearing his badge as police chief, sometimes stood at the intersection and blew his whistle when cars did not stop. Then he issued a ticket. It brought in a little money to the town treasury. People eventually got the idea that they had to stop at that sign.

As chairman of the board of selectman, Joseph was the official host for the Memorial Day celebration in 1936. That year, the first National Assembly of the Oxford Group took over Stockbridge. People came from twenty-four states and thirty-seven countries for the event. Members were housed in town, as well Lee and Lenox. During the next week, daily activities were held in each of the towns. Accommodating that many people in three localities took a lot of planning and seating. Luckily, the Berkshire Symphonic Festival Committee was moving that year from the Hanna Farm to Holmwood.[158] At Joseph's suggestion, they offered the benches to the Oxford Group. They voluntarily moved them to assembly sites in the three towns and returned them to the site at Holmwood when they were finished using them. Thus, money was saved for

---

[157] East Street

[158] Margaret Emerson's estate, formerly George Westinghouse's Erskine Park

both organizations. Joseph supervised the move because he was still in charge of the grounds for the festival and later had to move the shell as well. An estimated 5,000 people viewed the parade of 1,000 Oxford Group participants who marched down Main Street from Williams High School to the town hall. The parade marshall was United States Army General Erle D. Luce. A. Kensaton Twitchell and Charles Scoville Wishard from the Oxford Group followed. As chairman of the selectmen, Joseph represented the town of Stockbridge. Chief Uhm-Pa-Tuth, the last sachem of the Stockbridge Mohican Indians, dressed in buckskin clothes and a full eagle feather headdress, was with him. Next was the governor's Footguard Band from Hartford, Connecticut, the American Legion color guard, and the flag bearers of thirty-seven countries. There were many speeches at the town hall.

Dr. Buchman explained, "The aim of the Oxford Group is a supernatural network over live wires. A spiritual radiophone in every home with every last person, every last place, and every last situation in America guided by God. God-controlled super nationalism is the only sure foundation for world peace."

Others explained how the Oxford Group had changed their lives. Joseph's speech to the group, as quoted in the *Berkshire Evening Eagle*,[159] was very brief.

"By your teachings and conduct, you have left with us a deep impression. You have given us food for thought, needed so greatly in these most trying times. You have made history, not only for Stockbridge, but also for all of Berkshire County. The memory of you and your ideals will linger long in these hills. May your movement continue to gain momentum and spread throughout the world."

At the Indian monument and burial ground, Joseph received an eagle feather from Chief Uhm-Pa-Tuth in memory of his tribal connection to the town.[160] After they smoked a peace pipe, it was presented to the town.[161] Efforts of the Oxford Group, though of noble intent, did little to change the tide of history. War clouds were darkening in Europe and the Pacific Rim.

---

[159] See articles in *Berkshire Evening Eagle*, June 1, 1936.

[160] "Uhm-Pa-Tuth Lays Pipe of Peace on Burial Ground of His Ancestors," *Berkshire Evening Eagle*, June 8, 1936.

[161] These are in the historical collection at the Stockbridge Library.

Stockbridge was making its own mark in history when, on August 13, 1936, the Boston Symphony Orchestra made its festival debut. Three concerts were held at the estate of Margaret Emerson, previously owned by Westinghouse, on the western shore of Laurel Lake between Stockbridge and Lenox.

The exact site was in a large field in the southwest corner of the property. The same shell and benches as before were used with the addition of rented folding chairs in this new location. A tent seating 5,000 was erected over it. A smaller tent was provided for the food concessions. Everything else was arranged to handle a large number of people. Volunteers handled parking cars. Others were ushers and ticket takers.

These concerts brought even wider recognition to the world of music and foretold the prominent position the festival would occupy. The attendance was well over the capacity of the tent. Hundreds listened outside of the tent.

In 1937, the following year, Mrs. Gorham Brooks generously offered the well-known Tanglewood estate to the Boston Symphony Orchestra as a permanent home for the festival. The estate was where Hawthorne wrote *The Wonder Book* and *The House of Seven Gables*. It is a truly beautiful spot that is located on a bluff north of Lake Makeenac, which is still in Stockbridge Township. The serenity of the lake to the south contrasts with rugged hills extending all the way to Mount Everett.

The tent, shell, and benches were again used in 1937, along with the addition of another section. The tent had its charm, and the acoustics were good. Nevertheless, there was a lot more to the festival than music. Joseph had volunteered to take charge of the physical arrangements for the concerts. There was the tent and seating, plus three miles of electric wiring, thirty spotlights, three miles of fence around the parking areas, one-eighth of a mile of three-inch water main, four telephones, four Western Union telegraph wires for press and private messages, 800 benches, and 900 chairs. Personnel included 55 volunteer ushers and 50 volunteers for parking.

Six concerts were given. During one, a veritable cloudburst completely drowned the sound of the orchestra. Within the tent, there weren't any wet feet, but no one could get out while the furious patter of rain on the canvas roof made the orchestra inaudible. Dr. Koussevitzky made three attempts to proceed, but the Wagnerian storm was his successful rival. A concert hall was an obvious necessity to ensure the success of future festivals. Miss Gertrude Robinson Smith made an impassioned speech. Immediately, $20,000 was pledged. Machinery was set in motion to raise the balance.

Dr. Kousevitzky suggested his friend, Eliel Saarinen,[162] might serve as the architect. The board of trustees hired him to design a theatre for them. In the preliminary sketches he submitted, the cost of the building was around $300,000. This figure horrified the committee. Joseph was sent to ask him to compromise on his design to fit the modest sum of $80,000 that had been pledged by early winter.

"Raise $250,000, or do nothing at all. For $125,000, you could only build a shed." That was Eliel Saarinen's final word.

On December 15, 1937, Mr. Bentley W. Warren, president of the Boston Symphony Orchestra, wrote to the Berkshire Symphonic Festival, Inc.

"…(You) should select a successor to Eliel Saarinen. The latter's plans so far, as we have seen them, seem little more than elaborate sketches…"

In a letter dated December 20, 1937, the great architect resigned, providing the statement, "You can do what you want with my plans, but my name may not be mentioned in connection with any modification of my plans."

For this, he was paid his fee of $3,000.

When Joseph volunteered to design a building that would meet their budget of $80,000, the board of trustees held a meeting. Except for Messrs. Edman and Dwight, they voted to accept his offer. Joseph began work immediately to design a new building and prepare plans so that work could start on January 1, 1938.

---

[162] Famous Finnish architect (1873–1950)

Chapter Eighteen

The Shed

Even though Joseph had been accepted by the board of trustees of Berkshire Symphonic Festival as the new architect for a music pavilion and Miss Gertrude Robinson Smith was a solid supporter, a couple people on the board strongly did not approve. One of the two dissenters, Mr. Edman,[163] said it was a mistake. He felt another architect should be hired or they forego the entire project.

To those who did not know Joseph Franz, the idea of him designing a music pavilion may have sounded preposterous. However, he was an example of the designer/engineer who have historically built most of the world's edifices. Even into the early twentieth century, professional engineers were held in high esteem long before architects were recognized as professionals in the building industry.

Joseph's past successes in design and inventiveness in the electrical field or civil engineering field included revolutionary advances in the power industry. He had designed and constructed bridges, an observation tower, and commercial and residential buildings. Granted, the Music Shed was the largest building that Joseph designed and built, but the structural principles were similar, except for the one added element, specifically interior acoustics.

Much has been written about the Berkshire Music Festival for years, especially the Music Shed and its perfect acoustics in unique surroundings.

---

[163] Mr. Edman was the publicity Man for the Berkshire Symphonic Festival and Editor of the *Berkshire Evening Eagle*.

Figure 33—The Music Shed at Tanglewood

Thousands have flocked to Tanglewood from every state and many foreign countries to enjoy the music, superbly played by the Boston Symphony Orchestra. Nevertheless, little or nothing is known of the men who actually built it.

The Shed has been described as "a building of beautiful simplicity. It is the tent glorified."[164] When it opened in the summer of 1938, it was believed to be the world's largest and most unique structure, with natural acoustics, for symphonic music. It covers one-and-a-half acres of ground. In 1938, the fan-shaped structure had a seating capacity of 6,038.[165] If needed, about 3,000 more could be seated in the spacious colonnade. The outer curve is 404 feet in length, which is more than two New York City blocks.

---

[164] "Berkshire Festival" by Joseph Franz, Southwestern Musician, May-June Issue 1945

[165] The present seating is less because the original folding chairs have been replaced by wider, wooden-slatted wooden seats to accommodate overweight audience members.

When the board authorized Joseph to go ahead, he decided the site should be in the large field adjoining the Bullard estate instead of where the board and Boston Symphony Orchestra had suggested. He designed a new building, made the model, researched the new plans, and assumed full responsibility for the entire project.

Figure 34—Joseph's model of the Shed

The new plans were duly filed, carrying his name, with the Department of Public Safety in Boston. Working closely with Graves and Hemmes men, he approved the contracts for bids on the foundation, carpenter work, electric work, roofing, plumbing, sewage disposal, and grading.

In creating this wondrous venue, Joseph faced many personal frustrations and challenges. He fought with the board of directors as to the location and style of building. When he showed them his model, they could only see the roof and considered it a slice of pie. Some suggested the building should be more "New England" in character, like a big barn. Joseph felt, even the majority of the board, who accepted his plans, did not have a large amount of confidence in him. It was only in desperation that they went along with his plans because of the lack of funds.

When he let out bids for contractors, he chose Graves and Hemmes, a firm from Great Barrington. Mr. Edman, a board member from Pittsfield, insisted his friend Mr. Lindholm, a contractor from Pittsfield, was the only man capable to carry out this work.

Joseph countered, "If the lowest bidder, Graves and Hemmes, could not do the job, I know that I could."

Joseph would not be compromised by people with their own agenda for political or personal gain. This led to a great deal of adverse publicity in the *Berkshire Evening Eagle* and created many erroneous impressions.

The resignation of the Scandinavian architect or the fact that Joseph had taken over the job was not mentioned. He finally gave the information to the *Springfield Republican*.[166] This is the only published account with the correct facts. That caused more furor because the *Berkshire Evening Eagle* was not the first to get this story.

Perhaps there was some justification for keeping the resignation undercover because much of the money raised was with the understanding that the great architect was doing this work. Koussevitzky also had to be pacified. He believed everything was being carried out as originally indicated. It behooved Joseph to hold his peace, at least until the acoustics proved a success.

The Music Shed could have been a complete failure if Joseph had placed it where he first laid it out, that is, about thirty feet further back, closer to the bordering trees. He drove all of the stakes for the layout himself. When driving the rear stakes, he noticed a distinct echo from the woods. It disappeared as he went away from the front of the apex. As a result, the whole layout was moved about fifty feet so the stage was outside the echo range. If he had not had that annoying experience at the Hanna Estate with the first concert, Joseph might not have been aware of this.[167]

A fan shape is not unique in theatre design. It has been used since early Greek and Roman times as standard seating arrangement. However, in most contemporary theatres, the auditorium is enclosed in a rectangular building. The same building usually contains a large stage house with a fly gallery and

---

[166] "Music Shed Acoustics Said 'Perfect' After Test," *Springfield Republican*, June 13, 1938.

[167] To this day, you can get an echo if you clap your hands behind the tuning room building.

wing space for scenery storage and technical equipment for theatrical productions. The typical stage house costs as much or more than the auditorium.

The Berkshire Symphonic Festival was purely a musical event that had begun, like the Greeks and Romans, in the open air. Joseph's idea was to create a structure that would retain the feeling of the outdoors while protecting musicians and audience from the elements of unpredictable New England summers. A large stage house with a fly gallery was not needed, nor were exterior walls or corridors. What was needed was shelter for a large audience and symphony orchestra within an acoustic shell that could project the best possible sound.

The construction work was started on January 2, 1938. Many more headaches arrived. The piers on the southwest corner and several others had to be set on top of the ground. This was because a fill had to be made, which could not be completed until spring. The ground had to thaw out before the concrete was placed. The footing near the apex had to be dug about three feet lower to allow for excavation after the frost. When it did thaw, hay was piled over the entire ground and footings to keep out more frost. Joseph placed thermometers near the footings to check temperatures. If the ground had refrozen, thawing again would lower the level of the pier.

Grading could not be completed until May. Joseph tested the actual load-bearing capacity of the soil at footings. At the lower piers, near the apex, they hit water at four feet below ground level. Anybody experienced with hardpan[168] when wet will realize what this means. The whole mass acts similar to molasses. Borings showed rock about ten feet down. The cost of either piling or digging down to rock was too expensive, so stones were tamped down by hand until they had a bearing capacity of 3,000 pounds per square foot.

Continuing the work on the project was a lonely job. Only one of the directors, Mrs. John Lynch, and a few of the Stockbridge residents, including George de Gersdorff, came once to see the work in progress. Fortunately, all of Graves and Hemmes men cooperated well with the steel workers and each other. The purlins, spiking pieces, and beams were shipped one week ahead of the rest of the steel so the construction crew could assembled them.

---

[168] A layer of hard soil that is impervious to water

As the rest of the steel arrived, it was accompanied by more anxious times. It came by rail to the Lenox Station, but it was very difficult getting around the street corners to the site. The ground thawed, and the site became like soup. The derrick was set on platoons, and the steel had to be hauled in by large caterpillar tractors. That was some mess! Joseph tried saving the frozen ground by having the whole area covered with sawdust. This helped for the first few loads of steel, but, as the sun persisted, all of it gave way.

Figure 35—Initial steel at Tanglewood

The Bethlehem Steel people understood the problem. Somehow, the steel went up. The beams with the spiking pieces attached were raised as soon as the steel supports were erected, saving precious time.

Joseph drove the first rivet. There wasn't a ceremony, but pictures were taken.

Figure 36—Joseph driving the first rivet

At the time, steel construction was connected with hot rivets. The rivets were a special order so they would fit exactly into holes on the steel frame. The first delivery was the wrong size. Therefore, they had to be sent back and reordered. These were heated in a forge on the ground until red-hot. The smith took his tongs to pick up the hot rivet and tossed it to a man on the steel. That man caught the rivet in a kind of funnel that he quickly swung over to the hole in the steel. Two men with jackhammers pounded the rivet into the hole and flattened the ends, thus securing the connection.

Joseph's plans called for three posts in the center of the auditorium. They were designed to reduce the steel costs by allowing much lighter steel construction. The roof was entirely flat. A soundproofing material covered it so that even the most ferocious summer storm would not disturb the orchestra or audience under it.

Lumber for the roof and other parts of the building were sent by ship from Seattle to New York City. Fortunately, nonunion labor unloaded it. It was sent by truck just in time, before the union thugs confiscated the cargo.

The floor pitch was only three feet because Joseph felt that full vision was not essential to the enjoyment of good music. Lavatories and storage rooms were placed in the corner bays of the building.[169] Seating capacity was increased from the tent's 5,000 to 6,038. To steady the structure, steel cross rods were used for sway bracing in several bays.

Figure 37—Steel construction taking shape

---

[169] These have been removed in recent renovations.

Dressing rooms and a tuning room were in a building behind the apex of the Shed, but they were not attached to it. In the end, the only similarity to the first architectural design was a generic fan shape.

The biggest problem was filling a space forty feet high and one-and-a-third acres in area with sound. The requirements of a successful shell include the following. First, the musicians must be able to hear themselves so they may all play together at the same concert pitch. Second, the shell must act as a mixing bowl for all of the sounds and blend them together; and. Third, it must send the entire volume of sound out to the audience in increasing volume.

The shell in which the orchestra was to play was the same one that had been used since 1934. Made of three-eighth inch seasoned plywood supported on wood uprights and joists, it was very resonant. It had been wet and dried many times, so the grains of the wood and layers of plywood had opened, creating a mellowness to the music. One could feel the vibration of the shell by laying one's hand on the walls during a concert. There was no electronic amplification whatsoever.[170]

Acoustics were discussed with several people. Even Koussevitzky was consulted. He answered, "I do not know any thing about acoustics, except this much. Any hall with an audience of over 3,000 is no good."

Joseph showed the plans to Professor Fay of MIT as well as Professor Brown from Williamstown. Neither of them saw anything to change, but they said, if it did not turn out right, they could fix it. Professor Fay added the sound reflector to the back of the auditorium after the structure was complete. Slanting from the ceiling to the top of the open colonnade, it intensified the sound for the rear listeners. Joseph thought it was useless because the sound was just as good in the colonnade, outside the reflector's range, as well out on the lawn and 200 feet back, all the way to the original house on the property.

Because of the finite amount of funds, the shell and interior of the building were left unfinished. This was an unexpected blessing because the sounds from the resonant shell were sent into a maze of steel surfaces. These broke up the sound waves so well that there were no large reflecting areas to create

---

[170] In the early 1950s, the Boston Symphony Orchestra acquired control of the shed. They had the shell repainted with an oil-based paint. The original paint was water-based. Joseph was not consulted about their choice and would not have approved of the oil paint because it sealed the pores in the wood and destroyed the fantastic acoustics.

echoes with resulting dead spots. The exceptional acoustical qualities of the Shed were partially due to the economic constraints.[171]

Joseph found many ways to save both time and money. As an example, a 100-foot trench was needed for the electric cable to the shed. Steve Hopkins was in charge and hired five men to dig twenty feet apiece. The first man started at zero. The second started at twenty. The third started at forty. The fourth started at sixty. The fifth started at eighty. The trench was finished, and the cable was laid in one day Each man was paid for one day's work, regardless of when he finished.

In June, Koussevitzky and the Boston Symphony Directors came to Tanglewood, along with some of the festival board members to inspect the work. The roof was on. A bulldozer was completing excavation and fill. The site for the shell was not yet excavated. Koussevitzky, Judd Edman, and Joseph stood on the elevated ground, approximately the same height of the eventual stage.

Koussevitzky pointed to the steel. "What are you going to do with this?"

Joseph smiled and said, "For the present, we will have to leave it as is."

"Oh! This is terrible! Terrible!" he whined. "This looks just like a railroad station. Miss Smith wanted a shed. That's what it is."

"Now, Doctor." Joseph was trying to be diplomatic. "We don't have the money to dress it up yet."

"Ah!" he retorted, "If you did not have the money, you had no right to undertake this. You have a responsibility to the public!"

Mr. Edman chimed in, "Everybody must know that it is not finished."

No one knew how sick Joseph felt. If it had not been for his pride to complete the job, he would have resigned right there. All of them had a "tea" at the Curtis Hotel afterward. Koussevitzky frowned the entire time. He was insulted that he should be expected to perform in a railroad shed.

---

[171] In 1959, an acoustical canopy was installed to enhance and correct the damaged natural acoustics.

One month before concert time, the job was finished. The grounds were landscaped. Sewers and septic systems were installed, and the roads leading from the main highway into the festival grounds were built.

To test the acoustics, Joseph asked family friend, Roxanne Davenport, to sing onstage. She was a trained soprano and the wife of the West Stockbridge School principal. Her voice came across the vast space loud and clear, all the way to the back of the colonnade and beyond. Joseph breathed a huge sigh of relief.

At the first rehearsal, Koussevitzky gave the baton to Mr. Bergen and stepped off the podium. Together, he and Joseph walked all around the auditorium, to the farthest ends of the great arc.

"Ah! The acoustics are perfect!" he exclaimed. He turned to Joseph, "What do you think it is? The atmosphere?"

"Yes," Joseph agreed, "the atmosphere is very necessary."

Koussevitzky went back to the stage and told his musicians, "Way back in the corners, you can hear just as well as in front!"

The steel did not trouble him anymore. Moreover, it apparently has never concerned anyone else either.

If more money was raised to dress up the Shed, Joseph suggested wood roll curtains to close in the Shed at a cost of $12,000. Nevertheless, even that was never done. They would have been a permanent enclosure for the winter and could have been used in case of rainstorms in summer. Rain had blown in as far as the boxes, even though it never happened during a concert. The curtains could never have been entirely let down because they would have created a triple echo.[172]

When the Shed was finished, Joseph designed two souvenir items, a paperweight with the design of the shed on it and a pair of brass bookends with a bas-relief design of the shed. His son, Russell, was still working for the Norton Tool and Dye Company, so Joseph sought his advice on possible materials to use in making these items. Russell suggested brass for the bookends and pot metal for the paperweights. Joseph came to New York City to have the book-

---

[172] Wooden panels were built the next year, which close the Shed during the winter months, to protect the interior.

ends made, but he forged the paperweights in his own basement. He presented the items as gifts to Miss Smith and others. He sold some of them.

That first summer, one of the guest artists was a young soprano named Dorothy Maynor.[173] Koussevitsky believed in promoting talent whenever he found it. Miss Maynor had toured Europe. This was her debut with an American orchestra. Before she arrived in Stockbridge, there was a great controversy as to where she would stay because Miss Maynor was an African-American.

Stockbridge had always been generally a white, Christian town. It was proud of the Gunn family because their ancestors were the first free black people to settle in town. The family had owned property for many generations, and they took part in town affairs, like other citizens. They were not included in social activities of the cottage people, even though some were highly regarded as loyal servants. Most Stockbridge residents knew about the heroic Mumbet, a servant in the Sedgwick mansion. She saved the family's silver from rebellious ruffians when they invaded the town in 1787 near the end of Shays' Rebellion.[174]

African-American visitors in town could not stay at either Heaton Hall[175] or the Red Lion Inn. There was a "black only" boardinghouse on Park Street.

The Boston Symphony had booked Miss Maynor at the Red Lion Inn. When the management found out she was black, they canceled her reservation and booked her into the boardinghouse on Park Street. Koussevitzky was furious.

He called Gertrude Robinson Smith, who called Joseph.

"What are we to do?"

Joseph marched to the Inn and found the owner in his office.

"You can't expect a musical star or any celebrity to stay in second-class accommodations. If this woman is good enough to sing with Boston Symphony, then she is certainly good enough to stay at your hotel."

---

[173] "A Negro Singer From Virginia Makes an Exciting Concert Debut," *Life*, December 11, 1939. Ms. Maynor would much later establish the Harlem School of the Arts in New York City.

[174] Daniel Shays (1747–1825) was an American soldier and leader of a rebellion in 1786–1787.

[175] It was one of the two resort hotels owned by the Treadway family. It was torn down and replaced by housing for seniors.

Joseph further suggested that the inn did not want to have any kind of a scandal that might hurt business. The reservation was reinstated. Without knowing it, Dorothy Maynor made another kind of debut. She broke the color barrier in the town of Stockbridge.

When Koussevitsky realized the size of the audience for which he was playing,[176] he immediately asked for an increase of $1,000 per concert. He was only given $500 because the house was union-scaled.

Joseph then considered himself to have been a fool. He had taken on that entire job for a commission of a measly $1,000, which hardly paid his out-of-pocket expenses. He had lost ten pounds in weight, suffered many humiliations, and barely received thanks as a reward for all of his labor.

One letter dated June 18, 1938, from Mr. Robert K. Wheeler, a local realtor and board member, finally brought a bit of joy to his heart. It read as follows:

> Dear Joe,
>
> Professor and Mrs. Brown of Williamstown were in the office Thursday. They spoke in such enthusiastic terms about the Music Shed that I was more than ever anxious to see it. Last evening was the first opportunity I had. Mrs. Wheeler and I drove up and witnessed the beautiful sunset, as well as seeing the work you have accomplished. I want you to know that I think the job is perfect. First of all, you will recall that I was somewhat "on the fence" as regards to the location.
>
> Most fortunately, the present location was decided upon. It could not have been better. I think most of us questioned the appearance of the completed building, but we were delighted to find that, not only is it not an eyesore, it is an attractive addition to the property.

---

[176] Approximately 12,000 to 13,000

```
        As for the construction, I don't claim to be
        an expert but, from my observation, the work
        has been especially well done. I know the
        final result is due largely to your constant
        supervision. Please accept my congratula-
        tions. You tackled a big job, and no one could
        have done it better. I am delighted with the
        results, and I know everyone else will be.

        Sincerely yours,
        Robert K. Wheeler
```

The opinion of Mr. Wheeler was valued very much. He was one of the most successful businessmen in the entire Berkshire County.

Other letters came from several men he had worked with or consulted during the construction. Bently J. Warren of Williams College praised the acoustics.

Ralph Riddel from Bethlehem Steel Company said, "The steel went together like a puzzle. I have never seen it go together so well! The entire structure fit on the piers and was assembled without reaming[177] a single hole."

In March 1939, the *Architectural Forum* published the first article by Joseph about the building of the shed. Excellent photos of the construction were included.

Some years after the success of the Music Shed, Dr. Stella Owsley, editor of the *Southwestern Musician*, asked Joseph to write an article for that magazine. The "Berkshire Festival" was published in the 1945 May–June issue.

"Tanglewood's Music Shed is an Acoustical eccentric," another article Joseph wrote, was published[178] in New England Professional Engineer, January 1954.

Koussevitsky's friend continued to be mentioned in connection with the shed, even though he had doomed it to failure. Joseph bided his time. He was not interested in monetary reward because this had been a labor of love.

---

[177] A redrilling process that is used when two holes only partially align

[178] Much of the material in the last chapter and this one are taken from the above-mentioned articles.

However, he did expect thanks and the public recognition he deserved. It never came from the *Berkshire Evening Eagle* or the Boston Symphony Orchestra.[179]

In a letter to Miss Gertrude Robinson Smith in 1954, Joseph wrote, "I have no regrets when I see what can happen to buildings even when designed by great architects. I refer to the Concert Music Hall and the Chamber Concert Hall. Both buildings failed structurally."[180]

For many years, the only mention to the creator of the shed was a small, bronze plaque. It hung high on the center column of the inner colonnade and simply stated, "Joseph Franz, Engineer."

---

[179] Many years after he was deceased, full credit was finally given to Joseph Franz. Due to the long efforts of M. D. Morris, PE, and the Society of Professional Engineers, after the concert on July 9, 1967, his wife, Emilia R. Franz, son Peter Franz, and other members of his family accepted a plaque posthumously acknowledging Joseph Franz as the architect of the Music Shed at Tanglewood. The plaque was hung with the others that commemorated Gertrude Robinson Smith and others who had contributed to the establishment of a permanent Berkshire Music Festival. The plaque reads:

Dedicated to the Memory of JOSEPH FRANZ, P. E., who was responsible for the final design and construction supervision of this Music Shed

Presented February 25, 1967, by Berkshire County Chapter Massachusetts Society of Professional Engineers

All of these historic plaques have since disappeared due to the removal of the lavatories from the corners of the shed. Hopefully, they are in storage somewhere, awaiting an appropriate display area on the shed or the grounds. An exhibit, on loan from the Stockbridge Historic Association, was displayed in the main house for two seasons. It contained Mr. Franz's model for the Shed, his photo, and information about him.

[180] Both of these buildings at Tanglewood were designed by Eero Saarnen, son of Eliel, and were not coordinated in design with the shed.

Chapter Nineteen

The War Decade

Like most of the reading public in town, Emilia and Joseph were members of the Library Association. Sitting on the corner of Main and Elm Streets, the Stockbridge Library has been one of the centerpieces of the town since 1863. It is one of the earliest public libraries in America. The original edifice was the gift of J. Z. Goodrich and Nathan Jackson in 1863. It was a relatively small, square, gray, granite building on Main Street with a white, wooden portico over the front door.

Figure 38—Old Stockbridge library

In 1937, a gift from Miss Mary V. Bement and bequests from Joseph H. Choate and Arabella King made the addition of a new wing and central entrance connecting the two wings possible. These were constructed in 1937 and 1938. Miss Bement's generosity again provided for the exterior conversion of the old library to conform to the architectural style of the new additions.

Joseph was hired to remodel the old building and face the stone with brick to match the new additions. Work began early in 1940 and was completed by July 1. Always willing to do more than his share, he designed the new furniture for the children's reading room and donated part of his property for the Hoffman's Memory Garden.

Figure 39—Stockbridge library renovated and enlarged (with Franz home behind)

Working as a civil engineer, Joseph designed a new water system for Dr. Beatrice Berle's residence. The Berle's water supply consisted of a well dug about nineteen feet deep with the water level about fifteen feet down from the top of the ground. It was located to the south of the house between the barn and the road. Surface water drained into it. The cesspool was about ten feet from the well, which also drained into it. The water had a bad smell, and a plumber had hung a bag of charcoal into the water, supposedly to clear it. A small pump operated by pressure control was about three feet from the surface of the well. Joseph asked Mrs. Berle if they drank this water. The answer was affirmative. He took a sample and had it analyzed. It was completely contaminated. Joseph condemned this supply and found a good spring about one mile away up the mountain. He designed and supervised the building of their new water supply system.

Between these projects, Joseph took time for family trips. In 1940, a new Studebaker replaced the Franklin car. The first long trip was to Montreal in June for Russell's wedding.

**Figure 40—Russell Franz wedding.
Standing: Joseph, Emilia, and Russell Franz, Polly, Robert and Alice Hooper. Kneeling: Shirley, Peter, Joanna Franz, and Joan Hooper**

He had met his wife, Polly Hooper, on a ski slope. She was Canadian, and she was teaching physical education at the Oak Grove School[181] in Maine. Until their son Robert Peter was born, she occasionally taught there in the next year or so.

---

[181] It is now the Maine Criminal Justice Academy.

For a couple of years, Russell and Polly operated a summer camp in Maine. They borrowed money from Joseph to buy it. Unfortunately, they were forced to close the camp. They lost everything when World War II started. People did not have the gasoline to take their children that far, and families stayed closer together. After the war, Russell found a better paying job with the Ryder Corporation in Montreal. The family then settled in Canada.

In July 1940, Emilia drove to California via the southern route in the new Studebaker. Friends in Stockbridge thought she was crazy to go all that way alone with three young children, but Joseph knew she would be fine. The family enjoyed seeing many wonderful sights along the way and in and around Los Angeles. Once again, they stayed with Teddy and Joe Nolan.

Joseph flew out for the final two weeks of the vacation. It was the first time he had flown such a long distance, and he found it very exhilarating. After a couple more days in Los Angeles, the family headed east. Joseph did most of the driving. Their route back to the East Coast was via Boulder Dam. Because it was the largest power-generating plant in the world at the time, Joseph was naturally most interested in seeing it. They took the tour to the bottom of the dam to see the great turbines and hear about how the dam was built. The power industry had come a long way from where he had started.

Las Vegas was an unreal phenomenon for Emilia, who had been raised in conservative New England. She was amazed to find slot machines even in the bank, where she had gone to change some travelers' checks. There was gambling everywhere, not just in the new casinos.

En route to Bryce Canyon, they almost lost Peter when the back door of the car flew open. Peter was hanging onto the door handle and went flying out over the road. There were not any safety locks on automobiles yet. Joseph brought the car to a screeching halt as Peter dropped to the ground. Fortunately, no other cars were on the road. Except for a few minor scrapes and bruises, the youngster sustained only a good scare. It was more traumatic for his parents.

They spent one night in Ogden, Utah, after a stop in Salt Lake City, where they saw the Great Salt Lake and had a tour of the newly constructed Mormon Temple. On the way to Yellowstone Park, they spent the night in Warm River, Idaho, adding another state to their growing list of states visited.

Yellowstone Park was filled with awe-inspiring sights, from the famous Old Faithful geyser, which regularly performed every hour, to the thermal pools, mud geysers, weird stone formations, and yellow rock cliffs. The black bears were everywhere, and they didn't have any fear of people. When Joseph foolishly got out of the car to photograph the bears at closer range, he was followed back to the car. He could barely get the window rolled up before one bear had his paws on the side of the car and was trying to get in.

The long, dull drive through Wyoming, Nebraska, and Iowa then followed. They stopped in Chicago to see the Etting family for lunch. After a short visit, they proceeded through Illinois, Ohio, and Indiana. The last overnight stop was Buffalo so they could see Niagara Falls. By then, the children had seen so much of the spectacular scenery in the West that they were decidedly not impressed with the Niagara Falls. This was just a little water trickling over a few rocks.

That was the last time the whole family traveled together. In the uncertain time before the World War II, Joseph became increasingly aware of the importance of being self-sufficient and self-reliant. The Studebaker turned out to be a lemon because of its poor gas mileage and need for repairs in the first year. Joseph had planned to trade it in after two years of use. Fortuitously, he traded it in after only a year-and-a-half. It had survived the long trip across the country, and that was enough. The new car was a Plymouth, and it served the family until after the end of the war. When Joseph finally traded it in, it still had its original tires.

In January 1941, Joseph, Dr. Robert Hiden, and his wife, Tilly, organized the Stockbridge Ski Club. Winter sports were becoming more popular since Hannes Schneider had started his ski school in New Hampshire. Dr. Hiden had taken lessons from him and learned Schneider's Arlberg technique. Every Saturday morning until the end of March, about twenty children from six to twelve years of age, including the Franz family and Bobby Hiden, were taught the Arlberg technique on the hill behind the school.[182] In the three months, all of them became proficient skiers. In the following years, the ski club added many new members. Children's classes were continued as well as classes for adults. As skills improved, club members tried slopes that were more challenging, including Mt. Greylock, Catamount, and Beartown Mountain. It was a great comfort when Catamount installed the first rope tow in the Berkshires. Before that, one walked up the mountain. Cross-country skiing was also part of the club activities. Joseph later taught skiing for students at the Alteraz

---

[182] Berkshire Evening Eagle, March 31, 1941.

School in Great Barrington. By 1947, the ski club had thirty adult members. It was able to get a towline on Beartown Mountain. Joseph was elected president. Englebert Krichels, Harold Farr, and Mrs. John B. Swan were vice presidents.

Figure 41—Joseph cross country skiing

In March 1941, Joseph received correspondence from Europe. News from Austria and Czechoslovakia was beginning to sound increasingly ominous. The final letter from Rosa said she was suffering from angina and had become very depressed and pessimistic. She had become brainwashed into believing that Hitler and the Nazi movement were the salvation of Europe. She had alienated herself from the rest of the family, and she found fault with everyone. Her letter was so depressing that Joseph did not answer it. Ludwig and his wife rarely saw her anymore. On the other hand, Maria, who also had angina, still seemed to be doing very well with her business in Bratislava. She had many friends and seemed happy in Czechoslovakia, but she missed seeing her siblings.

Broadcasts of Adolf Hitler, heard on the radio, were becoming more frightening. Some have claimed that the best form of government may be a benevolent dictatorship, but, unfortunately, most dictators are easily corrupted by the power they hold. Certainly, this was the case with Hitler. When he marched into Austria, Joseph knew it would not be long before America would be pulled into another great conflict. News from the family in Europe once again ceased.

The horror of Pearl Harbor jarred America out of her isolationist cocoon. President Roosevelt had no choice except to declare war on Japan and the Axis powers. Conscription of young men was initiated, along with a system of rationing for essential materials and goods. There was little chance that Joseph would ever serve in the armed forces. He and Russell were too old, but there was his younger son. As it so happened, the actual war came earlier than anticipated, so Peter missed the fighting. He was inducted into the Army after peace had returned.

The family, like Americans everywhere, tried to understand and adjust to rationing. Along with gas and rubber, butter, sugar, and meat were also rationed. Joseph figured out how to supplement a couple of these supplies.

One way Joseph stretched the rationed gas was to turn off the motor and coast down all hills. He also never exceeded the forty-five mile per hour limit imposed by Massachusetts laws. His efforts produced an exceedingly good gas consumption per mile. Long-distance travel was accomplished by rail or bus, which still connected small towns and rural areas with urban centers.

To add to the meat rations, he bought twelve Belgian hares. The children were instructed that these were to be food. They were not to be pets. Only the stud rabbit could be considered a pet. He was named Thumper, and he lived a long life. Caring for the animals was relatively easy. Joseph built a large pen with a good-sized house at one end. It was on posts so the rabbits would be dry. He used a fine mesh wire for the floor of the pen and chicken wire for the other three sides. Food and excrement could fall through the mesh on the floor, keeping the pen clean. The mess on the ground could easily be raked up and thrown in the garbage. The children were taught how to feed and water the animals. When it was time to butcher the rabbits, they ran into a problem. Joseph could not kill anything. Emilia, though having been brought up on a working farm, could not do it either. The children were never even considered for the task. The shoemaker, Mr. Grande, whose shop was across the street, solved the problem. Once the hide was off the animals, Emilia could handle the carcass. No one had any problem eating the meat.

Learning how to stretch the meat supply meant learning new recipes as well. Goulash and spaghetti sauce became more frequent menu items. Cooking became a survival skill. Welsh rabbit[183] was an item included on the menu. Creamed tuna with peas and creamed chipped beef on toast replaced roast beef, a leg of lamb, or pork chops. Joseph had learned to cook when he lived by himself in New York City. Whenever his wife was away or ill, he had to do the cooking. He had two specialties: cream of spinach and Hungarian meatballs. After the spinach was cooked, he put it in a colander and pounded it until it was thoroughly mashed. Then he would add just a little butter and heavy cream. All of his children liked Joseph's cooking, and they soon learned to make the meatballs themselves. The recipe was simple enough: ground beef, bread crumbs, an egg, garlic, all herbs one wishes, salt, pepper, and lots of paprika.

Butter had a more simple solution. Milk for the Franz family came from the Highlawn Farm dairy. The Field family had a herd of prize Jersey cows that gave milk very rich in butterfat. It was delivered unhomogenized in glass bottles. Joseph invented a cream skimmer that consisted of a one-inch diameter chromium tube cut three inches long and rounded at the bottom. A six-inch handle with a square hook on the top was soldered on the side. A round hole was at the bottom of the tube. A stopper made of a round, steel ball on a long handle was inserted through an eye on the hook of the handle, which guided it to fit perfectly over the hole. The implement was inserted into the mouth of the bottle. The cream ran over the top, filling the container. When it was raised out of the bottle and held over a container, the stopper would be lifted, releasing the cream. This process would be repeated until most of the cream was skimmed off. When enough cream was saved in the refrigerator, it would be churned into butter.

---

[183] Also known as Welsh "rare bit." It is an English "savory" made of melted cheese in a bechamel sauce. It is served on toast or biscuits.

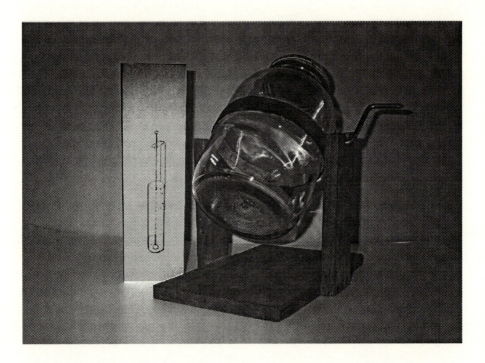

Figure 42—Joseph's butter churn & sketch of cream skimmer

The churn was another of Joseph's inventions. He took a gallon jar with a good screw-type top and fastened an iron strap around the center. To this, he welded two pins at the diameter of the circle. One of the pins was fashioned into a handle. It sat on a frame so it would rotate in a complete circle. As the jar was turned, the cream was gradually turned into butter as it separated from the buttermilk. The latter was used to make pancakes or waffles. Not only was honey or maple syrup used on pancakes and waffles, instead of the rationed sugar, it was also used in baked goods, on cereal, and in tea. Overall rationing never became a large problem.

Even though Joseph had never studied medicine, he was always interested in the subject. With interest, he read about new advancements in disease prevention and treatments for known illnesses. By 1930, studies had begun on the effects of diet in preventing ill health. One study was on the effects of calcium and the growth of healthy teeth and bones. The results sounded positive, so Joseph decided the family should start a regime to increase their calcium intake. He ground up soup bones to a consistency of cornmeal and sprinkled it on the morning cereal or used it in soups or in mashed potatoes. Years later, one could buy calcium tablets in the drugstore.

For a while, he also made tooth powder for the family. The basic ingredients for all tooth powder were simple enough: baking soda and salt with a little flavoring. Those were easily obtainable, and they cost far less than the drugstore products. It did not taste exceptionally good, but it did the job. Today, one cannot find anything except toothpaste on the market.

In Stockbridge, the children were progressing through school at a normal pace, but they sometimes needed a bit of prodding to improve their grades. Joseph tried helping with their math problems, but, when it came to algebra, he could never comprehend why they were taught to use fractions. It was such an inaccurate measurement. Emilia was good about checking their homework, book reports, and meetings with their teachers. Both parents attended PTA meetings when they could and took an active interest in school affairs.

Joseph and Emilia sent all three children to boarding schools for two to four of their high school years. Small towns just did not have the money to provide quality college preparatory education. School consolidation finally solved this problem.

Every year, the school held a Halloween party in the gymnasium. There were prizes for several categories of costumes. Emilia and Joseph worked together to outfit their children in some clever way. They once created a pumpkin costume for Peter. Another year, the girls were dressed as a Dutch boy and girl, complete with wooden shoes that Joseph had carved. A political cartoon was the theme another time. Each year, their children won prizes for the most original, the funniest, or the most unusual category.

Being a close-knit family, they played games together inside when the weather was rainy or snowing too hard to be outside. Monopoly was one of the favorites. Cards were a mental challenge as well as entertaining, particularly bridge and pinochle. They also listened to the radio. One of the first radios they had was an old crystal set that Joseph had built. They later bought models that provided better reception and more stations. The children loved the Sunday night programs of Charley McCarthy, Fred Allen, Jack Benny, and Fibber Magee and Molly.

During the week, after their homework was finished and chores completed, the children could listen to Jack Armstrong or The Lone Ranger. At five o'clock every evening, Joseph was in his favorite chair by the radiator in the living room, listening to the news program. The whole family gathered around for any of the news specials, which brought the outside world into their home. It is where they heard the horrible details of Pearl Harbor and America's involvement in World War II. It was also where they rejoiced on V-E Day and V-J Day when the war ended.

The periodic lessons in rudimentary survival skills continued. During the war, there were added lessons, specifically what to do in an air raid and how to escape a burning building. The room where Emilia stored her canned goods in the basement was stocked with additional tinned goods and water.

Joseph's personal habits were fairly simple. During his working years, he rose between 5:30 and 6:00 AM. He customarily had a light breakfast of a soft-boiled egg, toast or bread, fruit or juice, and coffee. On weekends, Emilia sometimes made pancakes or some kind of coffee cake with bacon or sausage. When he started working at home, the family always ate the main meal promptly at noon. That was partially important because Joseph had to get back to work and, later, the children had to return to school. Supper was always at six o'clock. In the winter, it was a light meal of soup or stew. In the summer, it was some kind of sandwich or maybe a salad. Meals were not always fancy, but they were wholesome. Because Joseph had a hole in his septum, he could not smell things, which affected his taste as well. It really did not matter what the food was. He ate everything. The family always complained that he ate too fast. One thing he could taste was any type of sweet. He always wanted something sweet to follow dinner and supper. His favorite desert was apple pie, which he invariably ordered when eating out, but he never refused anything else. In the 1940s, there were more Jell-O and fruit deserts than baked goods. Bedtime was generally at ten o'clock. He kept this routine all of his life, even when he was working on his own. The only exception occurred after he retired and he shifted his rising to the "late" hour of seven o'clock in the morning.

The vegetable garden in the summer became especially essential in the Great Depression years and, later, during World War II. Before he started work, Joseph spent at least an hour every morning in the garden. Two of the first things he planted in Stockbridge were asparagus and rhubarb. Both were productive every year and were the first things the family enjoyed from the garden each spring. He had also planted a grapevine that produced Concord grapes. In the Prohibition years, Joseph made wine, but it was a lot of work. Most of the time, Emilia made grape jelly and grape juice from the grapes. Raspberries were another crop that was relatively plentiful, depending on the invasions of Japanese beetles. Other crops varied from year to year, such as corn, tomatoes, beans, lettuce, and other types of vegetables.[184] The children loved climbing in the apple tree by the kitchen and never let Joseph trim any of the branches. The little pear tree always produced a good crop which Emilia canned or pickled. For a few years, a cherry tree was near the asparagus bed. Another apple tree was at the end of the back lawn. Joseph cut down the cherry tree after a couple years because the birds were getting all of the fruit and the family did not get any.[185] The second apple developed a blight, so it had to come down. Next, Joseph grew a Bartlett pear that bore too much fruit and became a mess. Peter planted a peach tree that finally produced two peaches. Both eventually succumbed to the saw.

Telling jokes was another form of amusement, particularly if there was a grain of truth in them. One of Joseph's favorites involved a local inebriate who liked to drink on Saturday nights and often missed Sunday Mass. The local priest constantly berated him for his errant ways. One day, the man was passing the time at Grande's Shoe Repair Shop, which was attached to the Van Deusen's Hardware Store on Elm Street. The local priest happened to come in.

"Good morning to ya, Father. Tell me, Father, what's the gout?" asked the man.

The priest answered, "The gout is it? Well, son, the gout is what you get when you go out carousing all night and don't atone for your sins on Sunday!"

The man laughed. "Ah, Father, I haven't got the gout. It says here in the paper that the Pope is suffering from the gout!"

While the original joke may have not involved anyone locally, Joseph always colored the joke to make it more interesting.

Joseph once told the children one joke with an extremely good moral. It was about the king's confidant who was sworn to secrecy. The king could talk out all of his troubles, but the confidant had to hold everything inside. One day, the confidant could not contain himself any longer. He went to the woods and dug a big hole. He hollered into the hole everything he had been holding inside. The moral was that "everyone needs someone to talk to."

---

[184] Two of the original fruit trees are still standing, a Baldwin apple tree near the back door and a sickle pear tree by the first garden plot.

[185] Maybe that's why young George Washington chopped his tree down?

Chapter Twenty

Jacob's Pillow

Joseph's final large project was his most rewarding. It was a unique theatre for the dance he designed and built at Jacob's Pillow.

Jacob's Pillow was an old farm, dating to circa 1790. The name came from a large boulder resting not far from where the house was built. It looked somewhat like a giant pillow. The farm, located just north of Route 20 in Becket, Massachusetts, is near the peak of the mountain that is called Jacob's Ladder.

Ted Shawn was an innovator in the dance world.[186] He had purchased the farm in 1930 to have a place where his company could spend the summers developing new dances for the coming year of touring. They apparently developed their muscles by converting the old barn into a studio, building small cottages for housing, and removing rocks to plant a garden of vegetables and flowers. In time, they also restored the main house, built a dining room, enlarged the barn, and repaired and maintained the outbuildings.[187]

Joseph's cousin, Caroline Fetzer, had studied dance with an Italian ballet mistress, Madame Elizabetha D. Menzeli, in New York City. Caroline performed in vaudeville, where she first met Ted Shawn and his wife, Ruth St. Denis. She joined the Ethel Gilmore Dance Company. For a few years, she toured with them. When touring became too exhausting, she stopped dancing and became a teacher. Meanwhile, her sister, Rose Marie, married and moved to Baltimore. Their father, Charles Fetzer, died in 1917. Shortly after, Rose Marie and her husband were killed in a horrible auto accident, leaving a small daughter, Virginia, better known as Boots. Carol and her mother subsequently moved to Baltimore to take care of the young child.

---

[186] Ted Shawn first became an innovator when he worked with his wife, Ruth St. Denis, and, later, when he formed an all-male dance company.

[187] See *How Beautiful Upon the Mountain* by Ted Shawn, published in 1943. A new history, *A Certain Place*, was written by Norton Owen in 1997. It was published by Jacob's Pillow Dance Festival, which Philip Morris funded.

Caroline changed her name to Carol Lynn for her career and was known by that name thereafter. She established her own ballet school and gained considerable recognition as "Baltimore's First Lady of Dance," becoming the head of the Peabody Institute's Dance Department. Through the years, she kept in touch with Ted and Miss Ruth. By 1934, she was coming up to the Berkshires in the summers to document Ted's work on film and help at the Pillow.

Joseph and the family had not seen much of the Fetzer family after they moved to Maryland, even though they did correspond. Once they reestablished contact, they never lost touch again. In 1943, Carol became dean of women at the Pillow, a post she held for several years during World War II.

In the summer of 1933, Shawn started giving Sunday afternoon "tea dances" to try out new material before an audience and help pay for some of the summer expenses. The programs were supplemented by Ted's educational lectures, as dancers and audience alike learned about the art of dance and its expressions of human emotions.

The barn was expanded to accommodate larger audiences in 1934. Performances were scheduled on Friday and Saturday as well as Sunday. Jacob's Pillow's reputation was spreading and the demand for tickets grew.

Carol invited Joseph and the family to see the work. They went many Sundays during the summer. Joseph dubbed these performances "old ladies' burlesque" because many portly ladies in the audience eagerly watched the scantily clad young men go through their routines. Following the program, tea was served in the garden as the dancers mingled with the audience.

The Franz children seemed to enjoy the programs as much as their parents did. Carol gave the children a few ballet lessons when she visited in Stockbridge and brought the girls secondhand ballet and toe shoes.

In 1941, Mrs. Norval H. Busey Jr. invited Joseph to become a board member of the Jacob's Pillow Festival Committee, a newly formed not-for-profit corporation. Her deceased husband had been on the Berkshire Symphonic Festival Board. Carol was delighted when Joseph accepted. The invitation also brought responsibilities, as Joseph soon learned. The first official meeting was on October 9, 1941, at Mrs. Busey's home. The committee was to raise $50,000. One-half of that cost was to go to Ted for the purchase of the property. The other half was to go toward a theatre. Joseph was elected treasurer.

Figure 43—Carol Lynn, Ted Shawn and Joseph at Jacob's Pillow

Ted had been impressed with the success of the Berkshire Music Festival and its ability to build a permanent home for music. The Shed was a crowning achievement. Ted had long desired such a solid establishment for the dance. In the summer of 1941, too shy to contact Joseph directly, he sent Frank Overlees[188] to ask Joseph if he would build a theatre at Jacob's Pillow.

When Joseph accepted the commission to plan the building, he asked Ted for his ideas on the size of the auditorium, number of seats, size of the stage, height of the proscenium, offstage space needed, number of dressing rooms, and, of course, the acoustics. Ted did not make any demands.

"Just build me a theatre," he directed.

He left the entire matter in Joseph's hands. Ted only stressed the necessity of having spectators looking down on the stage so they could understand the horizontal as well as vertical as movements of the performers and, at the same time, be aware of the intricate footwork. This is not achieved in most conventional theatres.

The theatre Joseph designed for Jacob's Pillow is a wooden structure. The original seating capacity was 625. The auditorium floor rises four-and-a-half inches every row of seats. The stage floor is three feet above the lowest level of the auditorium floor. It is made of wood resting on wood floor joists, giving it some flexibility to absorb the leaps and turns of dancers in motion.[189] Due to financial constraints, a fly gallery was deemed unnecessary. The proscenium arch is much lower than average, specifically sixteen feet above the stage floor. Only seventeen-and-a-half feet were needed from the arch to the ceiling line to accommodate the asbestos curtain, which was required at the time. That curtain had to "fly" in one sheet across the entire proscenium opening and fall of its own weight when released, just in case of fire.[190] This was always a problem in theatre design because it determined the height of the fly gallery and made a rather ugly outside architectural structure. None of this was visible at Jacob's Pillow. Instead, two small louvered gables protruding from the pitched roof accommodated the height of the fire curtain.

---

[188] Overlees was a former dancer with Shawn's troupe. He was then chief stage technician and stage manager.

[189] Both the auditorium and stage area were expanded in subsequent years of operation.

[190] In the 1940s, all public safety departments required this. Today, the curtain is no longer in use because it was replaced by a sprinkler system in 1990.

Figure 44—The Ted Shawn Theatre with weather vane of Barton Mumaw

Thanks to the suggestion of Frank Overlees,[191] lighting for the stage is accomplished by vertical rows of spotlights hung in concealed slots on the sidelines of the auditorium, invisible to the spectators. Additional spot and floodlights are hung from a beam near the roof of the auditorium. Other lighting instruments can be placed in the wings.

Scenery for the dance is greatly simplified or nonexistent in order to allow adequate space for the movement of the dancers, thus negating the need for a fly gallery or huge wing space. Unspoiled woods and a small pond surround the farm. Another of Frank's suggestions was to open the rear wall of the stage with barn-type doors to easily load and unload set pieces and other stage equipment. Never afraid to experiment, Joseph opened the entire back wall of the stage with sliding doors to accommodate the technical needs as well as use the natural scenery beyond as a backdrop. On a pleasant day, nothing is lovelier than a view of the Berkshire Hills.

---

[191] By that time, he and his wife, Patsy, had become good friends of the Franz family.

The theatre is built in barn construction and includes fifty-foot handhewn trusses. Warren Davis, a seventy-four-year-old African-American, who was one of the very few people still able to do this kind of work, personally chose and cut the trees on Lebanon Mountain. They were taken out of the woods by hand using a block and tackle because the terrain did not lend itself to either horses or tractors. They were hewn on the site.

Figure 45—Trusses hand hewn by Warren Davis

In addition to designing the theatre, Joseph made shrinkage tests of various woods. He learned that spruce, when dry, changes very little in overall dimensions. Instead, it opens up in the grain while pine shrinks all over. After the shrinkage tests, he decided to use a Shaker tradition in timber construction. He used all green lumber. The struts, or tenon members, are of spruce. The mortises are of native pine. This procedure has proved to work well. The Shakers used much the same principle, using both dry and green wood in the making of their chairs. Joseph searched extensively, but he could never find any literature on timber construction suggesting the use of the combination of spruce and pine woods.

Joseph Franz

The concrete foundations were poured in November 1941, just before America entered World War II. Construction began in April 1942. Snow was still on the ground. Because the building was erected of wood, there was no conflict with the wartime building restrictions.

Figure 46—Joseph and Ted Shawn viewing the progress of construction

Ted came up to see how the work was progressing. In the spring, he and some of his friends came to help clean the grounds and begin landscaping where possible. In early June, the theatre still did not have any sidewalls. By July, the exterior was finished.

The outside resembles the barn and former farm buildings. The roofline brings the building down to the landscape and is harmonious with other buildings in the complex. A louvered cupola for ventilation tops it. A steel plate weather vane that Joseph made in his basement crowns its pinnacle. In light of the Jacob's Pillow history, it was natural for him to choose a male dancer as the model for this weather vane.

"It is of Barton Mumaw, frozen in a pose from his dance in Bach's *Bourree*. He stands on a base, which portrays the rock, on which Jacob is supposed to have rested, dreaming of the ascension to heaven on Jacob's ladder. The figure is strutting out into the four corners of the earth as the will of the winds decree, always putting his best foot forward, not following the easy road, but struggling against the wind. It is also symbolic of any professional who is expecting to do the unconventional, creating new thoughts and ideas for which he must face the storm of public criticism until he, too, will find the long sought rest on Jacob's pillow rock at the end of his journeys."[192]

Everyone worked feverishly to have the building ready for the opening performance. Carpenters were still working on the dressing rooms as Agnes de Mille was rehearsing *Hell on Wheels* onstage. Students were picking up scraps of lumber and other building material from the ground. As a final addition, Joseph also made the antique-style lantern that hung over the front door.[193]

The night before the opening, Emilia spent the evening stenciling the numbers on the backs of the seats in the auditorium. On the opening day, Ted and Joseph were still tamping down a mix of mud and cinders in front of the entrance to the theatre. Some of the details were not finished until several weeks later, but the first performance continued as scheduled.

The first theatre of its kind, built exclusively for the dance, opened to rave notices. John Martin of the *New York Times* wrote the following in July 1942:

---

[192] Joseph Franz, "Dancing to the Four Winds," *The Berkshire Evening Eagle*, August 12, 1953.

[193] In 1990, the theatre acquired additional seating by renovating the original lobby area and changing the entrance.

```
An Ideal Theatre

The area is high and spacious, excellently
ventilated by louvers and redolent with the
odor of pinewood, yet, with all its native
color, it is a practical and up-to-date the-
atre, innocent of any taint of quaintness or
self-conscious rusticity. Everything about it
is designed for use with functionalism that is
both modern and in the frugal, straightfor-
ward New England tradition, classic. There
appears to be not a wasted nail or an unnec-
essary board anywhere, and the whole thing
operates smartly with a unity and efficiency
that are themselves a form of beauty.
```

The theatre cost $25,000. Joseph was to be paid five percent of the total cost for his services for designing and overseeing construction in the original agreement. He was actually paid $500 with the promise of the rest in due course.

Judging by the capacity audiences and large number of students, the festivals were always a complete success artistically, but never financially. The board of directors was expected to share any deficit. Joseph served seven years on the board. By 1947, he decided he was no longer fascinated with the dance world. He resigned, donating the balance due him to the Jacob's Pillow Festival, Inc.

Ted at least always acknowledged Joseph's contribution with genuine praise in his curtain speeches. He always said Joseph had the practicality to work within a budget and yet not compromise the artistic design. Joseph knew Ted was pleased with the theatre and appreciated his work. Publicity notices and historic references have also been generous in their praise. Joseph was extremely grateful to Jacob's Pillow, where he has always been given credit as the architect for the Ted Shawn Theatre and the Shed at Tanglewood. The sign attached to the front of the theatre reads:

Ted Shawn Theatre
Joseph Franz, Architect

All of his cultural endeavors were labors of love.

He once said, "We have to help the people who give us light and beauty and carry the world along."[194]

In 2002, the seventieth anniversary of Jacob's Pillow, the site was named to the National Register of Historic Places. On May 27, 2003, Jacob's Pillow was designated as a National Historic Landmark, as announced in Boston by Gail Norton, the Secretary of Interior. It is the first dance-related institution in America to receive this significant honor.[195]

---

[194] *Hartford Times*, July 11, 1953.

[195] Jeffrey Borak, "Jacob's Pillow wins designation as a National Historic Landmark," *The Berkshire Evening Eagle*, May 28, 2003.

Chapter Twenty-One

The Twilight Years

When peace was restored in Europe, the family again heard from Ludwig and Pepi, who were still living in Vienna. There had only been one bombing in Vienna, which mistakenly destroyed the opera house. The half-brothers had married and were living in Gumpolskirchen, along with Alma's daughter, Roselein. Thankfully, none of the family had been hurt. Recovery from the devastating effects of war was slow. The American family members once again sent food, clothes, and even a vacuum cleaner to Ludwig. Teddy sent food and clothes, particularly to the folks in Gumpolskirchen.

Survival of the Franz family members did not depend as much on external conditions. Instead, it was genetic or self-inflicted problems. Rosa committed suicide late in 1941. Heart trouble seemed to run in the family. Alma, the youngest, died in 1926. Eddy died in 1928. Karl had several attacks of angina before he succumbed in 1933. Angina plagued Marie most of her life, and her heart finally gave out in 1947.

The older generation of Joseph's relatives in America and Europe was diminishing. Karl Gumpinger was gone, as were Louise and her sister, Anna. Joseph had lost touch with most of his cousins during the Great Depression because their paths took off in very different directions. Joseph also suffered from angina. The first attack came when he was 47, and he had several recurrences. He fortunately kept quinine tablets nearby, and the spasms were of short duration.

His sister, Wetty, worked as a housekeeper for a wealthy family on Murry Hill in New York City after her husband died. Her daughters were both married. Aloysia had several children and lived in New York City. The older daughter, Katie, married a farmer and lived in Hannacroix, New York. They were close enough for Joseph's family to visit periodically. Katie had one daughter, Cathy.

Karl's wife, Mary, moved to Los Angeles to be near her daughter in 1944. Teddy and Joe were doing very well in California until Joe had a heart attack. He had been under terrible stress after Howard Hughes took over the studio. He recovered, but he could not continue at RKO at the same pace. Joseph and family saw them whenever they came east. Cousin Amalia Fetzer and daughter, Carol Lynn, visited every summer when they came to Jacob's Pillow.

Joseph's children were spread across the continent. Russell and his family lived in Montreal. Natalie and her family were in Alaska. The younger children matured and went to various colleges. In time, the girls married, moved away, and started families of their own. Peter finished college and military service and worked in Temiskaming, Canada.

It was easier to travel after the wartime restrictions were lifted. The family made one more trip to California when Joanna graduated from the University of Southern California in Los Angeles. This time, they drove out in the Plymouth. Once again, they stayed with Teddy and Joe at their new home in Upland. It was a wonderful place built in an orange grove. It had a swimming pool in the backyard. Every morning, they had fresh oranges off the trees for breakfast.

Joseph made a couple trips alone to visit his children. One winter, he went to Canada to see Peter, who was working for a paper company. The town had a nice skating rink, so he took along his ice skates. He could not resist teaching some of the youngsters a little of the finer points of figure skating.

He made another trip to California in 1952. By then, airplane travel was much more common. He wanted to see his daughters, who were living in San Diego then. Shirley already had two children, so it was an opportunity to see his grandchildren. He did stop in Los Angeles to see his niece and her husband for one day en route to Alaska to visit Natalie and her family. True to his nature, he never gave much advance warning for his visits. He simply appeared one afternoon in Homer, where Natalie and her family had settled. Natalie was disappointed she could not arrange more for him to do in Alaska, but there simply was not enough time. He only stayed a couple days and was off again.

Conducting educational lectures enlightened not only young audiences but also adults, like Einstein's theory of relativity, which he presented for the John Sargent Society in Stockbridge. One of his favorite subjects was the quality of education.

In 1946, 1947, and 1950, Joseph was asked to teach figure skating for the girls at the Oak Grove School, where Russell's wife had taught and from which Joanna had graduated. He spent a couple weeks each winter helping them with a Winter Carnival and giving several scientific lectures.

Figure 47—Joseph giving a lecture at Oak Grove School

At sixty-five, Joseph finally gave up riding and skiing, but he continued ice-skating. During the winter months,[196] Joseph volunteered his time to teach figure skating for the Pittsfield Parks and Recreation Department and in Sheffield. Some of the youth were unsteady on their feet, and the outdoor rinks did not have any handrails to help them maintain their balance. Joseph invented a support, similar to the one he had made years earlier for Natalie. It could be pushed along as the young skaters made their way on the ice.

---

[196] *Berkshire Evening Eagle,* January 24 and 29, 1995; February 2, 1955.

**Figure 48—Joseph's ice-skating support
(Berkshire Eagle photo by Fowler, January 15, 1957)**

He was ice skating until six months before he died!

**Figure 49—Joseph still ice-skating
(Photo courtesy of the Stockbridge Library Association's Historic Collection)**

Ironically, his brother, Ludwig, died in 1954 when he fell through the ice while skating on the Danube.

Realizing his limited ability, for his own self-preservation, he began resigning from some of the organizations to which he had belonged. First, it was from the Laurel Hill Association in 1943. In 1948, he resigned from the Trustees of Reservations. In 1957, he left the Society of Military Engineers, even though he had a life membership. The one organization that still gave him a great deal of pleasure was the AIEE. He never missed a meeting.

Figure 50—AIEE meeting, Joseph with John Alberti from General Electric in Springfield, Massachusetts

Joseph remained active with small-paying jobs that were still coming his way. The Congregational church wanted to expand their parsonage in 1948 and called Joseph to survey their needs. He did so and made the plans for the new addition. The Hurlbut Paper Company in South Lee hired Joseph to design and construct an elevator shaft as an addition to their mill. Along with that, he made the seal of their top hat logo for them.

The Berkshire Playhouse was originally known as the Stockbridge Casino. In 1927, it was moved from Sargent Street to its present location on East Main Street and Yale Hill and remodeled into a theatre. It has been a popular professional summer theatre ever since. One alteration in 1937 added an orchestra pit. By 1953, the Three Arts Society, which owned the building, decided more seats were needed. They hired Joseph to draw up plans for the additional seating. He added forty-five more seats with a cantilevered extension of the balcony. This added two more rows. By remodeling the basement stairwell, he gained two more seats on the main floor, a total of forty-seven seats. Other alterations included a new ventilation system, enlargement of the lobby and upstairs lounge, and removal of the women's restroom to the basement. Without disturbing the original McKim, Mead, and White[197] architecture outside, the renovations were completed in time for the opening of the 1954 season.

Additionally, he continued volunteering his time for community projects. Joseph served on a committee in 1950 to survey the feasibility of a water system for the Furnace District. He presented three possibilities at a special town meeting, but the town voted to investigate the matter further. The Garden Center needed an addition to the original house. Joseph designed a garage for the staff between the kitchen and the shed with more offices behind it. As was typical, he also assumed responsibility for supervising the construction. Then he designed a bowling alley for the Stockbridge Community Service Organization.

Joseph continued to be outspoken on questions of education, taxation, and even sonic booms. He wrote numerous letters to editors of the local papers on these subjects. Because of his concern for the quality of life in Stockbridge, he was appointed to the Stockbridge Board of Appeals for a three-year term and was elected chairman. The board held hearings at the town office for all kinds of issues, like rezoning someone's property when they wished to operate a business on their premises. Still concerned with his town's affairs, Joseph ran

---

[197] Berkshire Evening Eagle, January 15, 1957.

for town moderator in 1953. Heaton I. Treadway, the incumbent and son of the deceased congressman, defeated him. The town hired him that year to estimate the cost of restoring the Gideon Smith Cemetery in Interlaken. The board of selectmen used his estimate to present the question of restoration at the town meeting.

By 1955, Joseph's family was even more widely scattered. Russell and Peter were in Canada. Natalie was still in Alaska, even though her husband had died at very young age of a massive heart attack. Their daughter, Anna, was at school in Eastern Massachusetts. Shirley and her family had moved to Arlington, Virginia. In addition, Joanna was living in the Middle East with her husband, where they had made many friends.

That Christmas, Joseph and Emilia had two guests from Iran and one from Libya for the holiday season. Joanna's husband, John Humphrey,[198] had worked with three young men in their homelands, and they became friends. The three were spending a year's special training in film production at Syracuse University.[199] During the spring of 1956, they had more foreign visitors, including another friend of Joanna and John from Iran, Dr. Mohammed Taqi Mostafavi, and his interpreter, Dr. Elio Gianturco, who was from Italy. Dr. Mostafavi was the director of the Tehran Archeological Museum. He had been a technical director for an educational film, *Views of Iran*, that Mr. Humphrey had made for the Shah under the United States Foreign Aid Program. He was to give a lecture at the Berkshire Museum in Pittsfield as part of a three-month educational exchange tour. The United States State Department booked them into the Oaklawn Inn, which had been closed for more than ten years. Joseph and Emilia welcomed them to stay in their home instead. They also gave them a tour of the Southern Berkshire area.

There were several more holiday reunions with Joseph and his family, particularly after 1957. One last time in 1959, to help celebrate Easter weekend, Joseph had all of his children and a few grandchildren with him.

After several years of enervating heart disease, Joseph Franz died quietly at home on Tuesday, June 23, 1959.[200] He was just three weeks shy of seventy-seven. He lived long enough to be reunited with his favorite sister, Wetty, who

---

[198] The Humphreys lived in Iran until the fall of 1954. Then, they lived in Libya until 1955 and lived in Iraq until 1958.

[199] Berkshire Evening Eagle, December 15, 1955.

[200] Berkshire Evening Eagle, June 23, 1959.

came to live with Emilia and him during the last few months of his life. He saw all of his children married, except for Peter. All were living productive lives. He was also survived by his half-brothers, Karl and Otto; seven grandchildren; and one great-grandchild.

His most vigorous, unresolved fight had been with the many articles from various publications, which downplayed his role in the creation of the Music Shed at Tanglewood with incorrect information. His wife and family continued the war long after his death. For the celebration of the fiftieth year of the Shed, Emilia, Peter, and Peter's wife, Deborah, were special guests of the Boston Symphony Orchestra, along with surviving members and relatives of the original Berkshire Symphonic Festival Trustees. The recorded speeches from the opening concert in the Shed were played, and Joseph Franz was acknowledged as the builder.

Nevertheless, it was not until August 24, 2003, when the sixty-fifth year of the Shed was celebrated, that he was acknowledged from the stage as the architect and builder. For that celebration, his daughters, Joanna and Shirley; granddaughter, Deryl Clune; and great-granddaughter, Aliza, were guests of the Boston Symphony, along with current trustees and administrators. Unfortunately, the printed material in the exhibit and program still referred to Joseph as having modified the Finnish architect's designs. The goal is in sight, but there are battles yet to be won.

Joseph left behind his immortality. He helped brighten the world with his skill and electrical inventiveness. As a worker, he was praised by fellow professionals as "thorough, capable, and efficient in the field of electrical engineering." He was "at all times a thoroughly reliable worker and efficient" with "a quick insight into matters electrical and [was] a progressive engineer."

His love of the arts and his engineering prowess made his architectural designs a reality. These have brought—and continue to bring—many people to the Berkshires, which have become a summer cultural capital.

He freely donated his time to preserve the natural beauty in the town of Stockbridge and the surrounding hills. As a concerned citizen, he served his town in various political posts, elected and appointed.

Miss Gertrude Robinson Smith once wrote in the *Berkshire Evening Eagle*, "We are particularly fortunate in having had Joseph Franz as chairman of the selectmen. His training has been that of an engineer, and he is a thorough businessman. He is a man of integrity with no axes to grind, who assumed his duties toward the town at a personal sacrifice."

His contributions to Berkshire County included service on the board of directors of cultural and service organizations. Whether lecturing to schoolchildren or sharing his lifelong passion for ice-skating and other sports with the community at-large, he brought people of all ages and lifestyles together.

The following letter to the *Berkshire Evening Eagle* editor appeared on April 10, 1958. It best describes Joseph's humanitarian persona and was written by William A. McCarty from Housatonic.

> Concerning Joe Franz
>
> I enjoyed reading Anthony Rud's column [Our Berkshires, April 2] very much. It called attention to the many outstanding things that Mr. Joseph Franz has done in our community, and, by that, I mean all Western Massachusetts. I have known Mr. Franz for about fifty years and am familiar with his many deeds of service over the years that have passed. What has impressed me greatly is his training of young men in electrical work. Joe has always been a natural teacher, and there are scores of men who first learned about electricity from him.
>
> While the Franz projects named by Mr. Rud were outstanding, to me, his most thoughtful was installing fixtures of a complex nature at the spring by the side of the road below the Glendale powerhouse so that passersby can partake of the wonderful water that flows so freely there. Over the years, thousands of people in the neighborhood come with jugs and pails to this piped spring to take this wonderful water to their homes. In some instances, people come to this spring from many miles away.

As a father, he was caring, kind, and a strict, but gentle, disciplinarian. He was a patient teacher with a flexible point of view and was an example of unprejudiced ideals and principles.

Few in the area have given more of themselves than Joseph Franz had. He did so without expecting financial rewards. His only wish was to be accepted and appreciated for his accomplishments.

Henry Wadsworth Longfellow's poem "A Psalm of Life" impressed him greatly in his early years. As he once said, "it has been a guide and comfort all of my life." It seems a good place to end this book.

Tell me not, in mournful numbers,
Life is but an empty dream!
For the soul is dead that slumbers,
And things are not what they seem.

Life is real! Life is earnest
And the grave is not its goal;
Dust thou art, to dust returnest,
Was not spoken of the soul.

Not enjoyment and not sorrow,
Is our destined end or way;
But to act that each tomorrow
Find us farther than today.

Art is long and Time is fleeting
And our hearts, though stout and brave,
Still, like muffled drums, are beating
Funeral marches to the grave.

On the world's broad field of battle,
On the bivouac of Life,
Be not like dumb, driven cattle!
Be a hero in the strife!

Trust no Future, how'er pleasant!
Let the dead Past bury its dead!
Act—act in the living Present!
Heart within and God o'erhead!

Lives of great men all remind us
We can make our lives sublime,
And departing, leave behind us
Footprints on the sands of time.

Footprints, that perhaps another,
Sailing o'er life's stormy main,
A forlorn and shipwrecked brother
Seeing, shall take heart again.

Let us then, be up and doing,
With a heart for any fate;
Still achieving, still pursuing,
Learn to labor and to wait.

978-0-595-36046-8
0-595-36046-7

Printed in the United States
200288BV00005B/217-234/A